Small Flying Drones

Gianluca Casagrande • András Sik
Gergely Szabó
Editors

Small Flying Drones

Applications for Geographic Observation

Springer

Editors
Gianluca Casagrande
Geographic Research and Application Laboratory
European University of Rome
Rome, Italy

Gergely Szabó
Department of Physical Geography and Geoinformatics
University of Debrecen
Debrecen, Hungary

András Sik
Lechner Tudásközpont
Budapest, Hungary

ISBN 978-3-319-66576-4 ISBN 978-3-319-66577-1 (eBook)
DOI 10.1007/978-3-319-66577-1

Library of Congress Control Number: 2017955957

Printed on acid-free paper

This Springer imprint is published by Springer Nature
The registered company is Springer International Publishing AG
The registered company address is: Gewerbestrasse 11, 6330 Cham, Switzerland

Foreword

Small civilian drones are commonly used in various professional sectors. Obviously, they are also raising interest in geography and related contexts. An old geographer saying states that our work must be done "on foot," in reference to the importance of field work. Going to places to observe and acquire data is still one of the best ways to do research and develop applications. In recent decades, however, we have discovered the usefulness of new tools and are now able to "capture" and "record" phenomena which could not easily – or conveniently – be detected by traditional techniques. In some cases, these tools have the capability of discovering otherwise invisible elements. In many other cases, they allow us to see things more effectively. Both aspects are important. The first one advances the human ability to discover; the second fosters innovation of methods, an important element of scientific research and socioeconomic development.

Over the past few decades, for some types of investigations, geographers have looked at the Earth's material aspects using aircrafts and, more recently, satellites. This has remarkably enhanced our knowledge about the environment and anthropized spaces. New concepts, processes, and techniques were developed in this field; they were progressively integrated with geographic information science and became a widely shared patrimony, and thus new approaches and methods were added to the geographer's toolbox. Today, new opportunities for data acquisition, elaboration, and presentation are available, along with a new set of remote sensing technologies. Small civilian drones have enabled far more users to produce their own "views from above" without the need to obtain them from institutional or commercial providers. Like in the case of other technologies, possibilities for drones will be developed through practical use. Much is already available, and more will surely come in the future.

Meanwhile, in different countries, technologies and regulations evolve in search for the necessary – and sometimes difficult – balance between operational needs and juridical limits. This book includes reflections on the potential of small drones in several fields. The authors and editors of the following chapters are experts in geography, remote sensing, and architecture. Hence they are representatives of high interest categories in the use of these new instruments.

The first three chapters serve as a general introduction to the topic; they present the reader with a wide picture of opportunities and issues about the new technologies and their use. Attention is paid to practical and legal aspects. The fourth chapter provides a review of the different types of instruments which can be used for remote sensing and geographic observation. Finally, the fifth chapter emphasizes specific applications. Some parts of the text go deep into technological, bureaucratical, or operational discussions. At first glance, this might seem out of the expected scope of a publication about geography. However, it would be impossible to use small UAVs (*unmanned aerial vehicles*) without taking those aspects into proper account. It is then recommended to give the same attention to all parts of the volume.

The book warns the reader not to overestimate the opportunities brought about by the new technologies. Some inherent limitations of mini and micro-drones are presented as well. In our opinion, the most important caveat about UAVs is that in spite of automated survey and analysis algorithms, small drones remain able, per se, to acquire *data*, not *knowledge*. The latter is still – and will remain in the future – the result of an interpretation which can be fully developed only by the human mind. Once this point is established, then drones can be considered powerful and very innovative tools. They prove flexible and easily adaptable to different needs: the needs of many geographers and other professionals involved in studying the environment, landscape, and territory, as well as some cultural and social phenomena.

After reading this book, a conceptual synthesis becomes simple. For the type of information they can collect and for the effectiveness of the representations they create, drones are already highly valuable technical aids for meaningful geographic research and work. They will be even more so in the future.

Rome, Italy

Prof. Filippo Bencardino
President of the Italian Geographical Society

Preface

The increasingly massive spread of small drones in the general market accompanies a rapid evolution of payload hardware and software. From photo/video to multispectral sensors, to environmental measurement devices, up to mechanical operation peripherals, even basic drones are getting more sophisticated and miniaturized at the same time. The recent advent of minimal-sized professional drones is specifically to be taken into account, as it can make a substantial difference in terms of usability.

There is some variability about the technical name assigned to these machines. This is generally due to the emphasis put on specific aspects of their nature or goals. The most widespread name of "UAV" (unmanned aerial vehicle) is very general and could apply to any machine able to fly without a crew; the focus in this case is on the aircraft itself. "UAS" (unmanned aerial system) generally implies that the vehicle is part of a wider system, involving a ground control station and possibly other components; "RPV" (remotely piloted vehicle), "RPA" (remotely piloted aircraft), and "RPAS" (remotely piloted aerial system), and their national language variants, emphasize that the vehicle is actively piloted from a remote station.

From a strictly ontological point of view, a difference is admitted between UAVs, in general, and RPAS. The latter are "remotely piloted" while the former could be any kind of aerial vehicle without an onboard crew. This includes then the types of platform which can be taken into the air but with no control whatsoever by a ground crew: free-flight balloons and airplanes, some types of rockets, etc. The distinction is substantial, but current use tends to converge on "UAV" in scientific contexts and "drones" among laymen, so we prevailingly stick to these two names.

Obvious applications can be found, from basic cartography and photogrammetry to specific fields of environmental studies and physical geography. The data acquisition capabilities of drones can also be applied to phenomena which are interpretable in terms of human geography. From "thinking" of a specific survey to actually "doing" it, however, a lot of nuts and bolts are to be put into place.

Meanwhile, aviation authorities all over the world are dealing with the complex task of regulating a set of technologies whose fast development often outruns the ability of setting rules for them. Once a regulation is issued, chances are it was developed around a specific concept of drones; but it frequently turns out that drone technology has already gone farther. This way, the recently created rules tend to freeze technological development at a certain stage, hampering – de facto – the full development of new tools in many fields.

We are experiencing the actual "dawn of the drones," and their application potential is boosting everyday. If UAVs are to be acquired as long-living standard tools in geography-related investigations or applications, then it is worthless to get too deep into current implementations or regulatory stages. Rather, concepts, opportunities, and issues are to be considered in a more general sense.

This book does not try to establish recommended solutions for all cases; it specifically introduces concepts about flying small drones (not *any* drone) in some fields (not *any* field) of geographic and geography-related observation. It will therefore take into account questions, answers, and perspectives that directly involve operators as they plan their work before they get to the field.

Rome, Italy, and Budapest, Hungary The Editors

Contents

Chapter 1
Small Drones and Geographic Observation

Gianluca Casagrande

Abstract This is an introductory chapter in which the overall scenario of the book is presented. Theoretical aspects are discussed so as to clarify what is the interest of using small drones in geographic observation (1.1), what the most common application scenarios and limitations are (1.2), and some general perspectives about future developments (1.3).

Keywords Small drones • UAV • Aerial view • Landscape • Territorialization • Reification • Geographic observation

1.1 Getting the Overview

This section introduces some basic concepts of UAV-based geographic observation from an epistemological point of view. The purpose is to demonstrate how drones, as technological tools, may enable geographers to more efficiently perform a series of tasks. Among these are different kinds of analyses, whether in hard-science or social science/humanities-related themes. New tools are showing, at the moment, a potential of similar usefulness in both the so-called two "worlds" or "grammars" of geography (Buttimer 1993, p. 215; Vallega 2004). If the reader wishes to exclusively focus on technical or application-related topics, we suggest to skip this section and proceed to the following chapters.

Geographic observation includes, but generally goes beyond, the basic study of physical environment. It actually aims to comprehend spaces, territories, and cultural landscapes in their material dimensions and also in some of their immaterial ones. This is possible inasmuch as the latter can be revealed or indicated by the configuration of material elements. The basic idea is that actions by human individuals and

G. Casagrande (✉)
Geographic Research and Application Laboratory, European University of Rome, Rome, Italy
e-mail: gianluca.casagrande@unier.it

G. Casagrande et al. (eds.), *Small Flying Drones: Applications for Geographic Observation*, DOI 10.1007/978-3-319-66577-1_1

groups in their *espace vécu* (Frémont 1976) are the result of choices derived – often impredictably – from a combination of material and immaterial factors.

Such complexity could be read by taking into account several epistemological and methodological approaches (Buttimer 1993, p. 216–221). By using drones – as any other survey tool – we can effectively investigate aspects of the physical space. In our analyses, such space can obviously appear as a physical *environment* or rather as an instance of that particular mixture of natural and artificial space that is called a cultural *landscape* (Haggett 2001, p. 223–226, 336; Turco 2002, pp. 7–8).

Anthropization of a certain site or region is induced (but not determined) by local conditions, on the one hand, and by interactions with the surrounding anthropized areas, at different scales, on the other hand.

The area's inner functional system is to be integrated with lower-level, peer-level, and higher-level systems. In the final analysis, however, people and communities first budget the aforementioned conditions and then drive their own *territorialization* process (Raffestin 1977, 1982; Turco 1988, pp. 74–134, 2002, pp. 9–10). By doing so, they establish and develop economically and culturally shaped spaces and networks of systems.

An important theoretical distinction was set, in geography, by the terms "space" and "place" (Tuan 1977). The former is a portion of Earth's surface that can be defined through sets of spatiotemporal coordinates; the latter can be understood as physical space(s) endowed, by the people who are in it, with experiential, existential, and symbolic values. Naturally, the symbolical texture of a certain space depends on the identity of the observer, whether the observer is a specific individual (drawing from her/his experience) or a member of a community referring to a specific shared system of heritage and values (Bignante 2010). Building from Vallega (2003) it becomes evident that the evolution of anthropized spaces can be expressed as a general feedback iteration among three mutually influencing dynamic nodes: perception, material transformation (also called "reification"), and regulatory organization.

Through *time*, each node in the process evolves due to its inner mechanisms and also by influence of the actions and reactions of the other two nodes.

In summary, as it was stated before, human groups shape their places by adapting them to material conditions; nevertheless, this adaptation occurs based on perceptions and intentions of actors and decision-makers. The process takes place in relation – not necessarily in agreement – to formal or informal regulations. In this overall mechanism, culture is a major driver (Guarrasi 1992, p. 32).

In principle, we could observe how a certain space is configured from environmental and reification points of view. We could then use the acquired information to grasp knowledge not only on material and (possibly) regulatory status of the area but also, at least in some degree, about decisions and perceptions by communities in their living space.

As geographers observe spaces and places from above, regardless of the means they deploy for this purpose, phenomena can be analyzed according to at least three different levels of interpretation: material, functional, and symbolic.

At the material level, nature, location, and purpose of specific elements can be analyzed. This basic type of interpretation is highly developed and widely applied in remote sensing. Analyses are based on the fact that objects can be identified, measured, and examined in their physical nature.

At the functional level, material features of landscape are observed to interpret processes, flows, networks, and services. Analyses are conducted by recognizing – either directly or indirectly – status changes in different types of spaces, operational conditions of facilities and infrastructures, transitions in the availability of resources, usage of services, and variations in "footprints" produced as a side effect of human activity (Figs. 1.1 and 1.2).

At the symbolic level, landscape is surveyed in its material features in the attempt to identify elements of the presence and/or actions of a specific culture, whether this presence is current or past. In this case, analyses are oriented to recognizing specific material "markers" which can be associated to one or more systems of values, beliefs, and identities (Fig. 1.3). From this point of view, drone observation can be thought of as a twofold knowledge acquisition means. Indeed, on the one hand, it can provide information and data about symbolical markers (also called "referents") present in the observed scene. On the other hand, it is the drone itself that, by being guided so as to acquire data about a certain landscape, may provide information on priorities and perceptions of the users in *reading their own* landscape and cooperating to its observation (Turco 2010, pp. 263–266; Bignante 2010; Grainger 2017).

Considering the three mentioned levels of analysis, it is worth underlining the importance of time as the fourth fundamental coordinate of any geographical study.

The ability to monitor change in spaces and places through time is important in all fields of applied geography, from basic remote sensing to complex evolutionary analyses. Many investigations would be of poor or no use unless they could be based on the comparison among different epochs in which the status of a certain variable is acquired. Data acquisition, depending on the observed phenomenon, could require a more or less regular periodization.

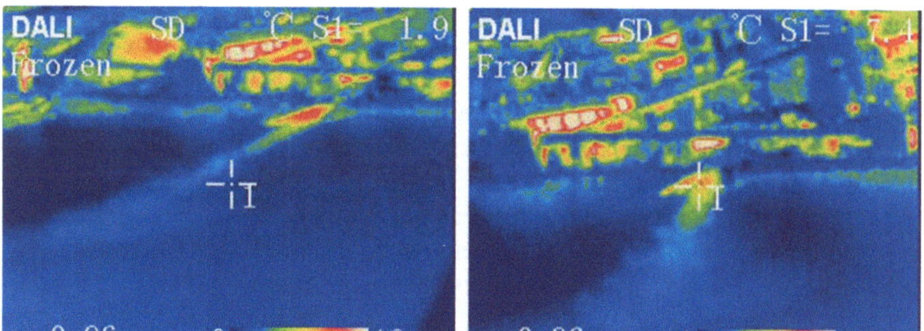

Figs. 1.1 and 1.2 Tracking human activity in the environment. Low-resolution thermal infrared images of the cooling water outlet of an electric power plant into the Tyrrhenian Sea. A plume of warm water is clearly recognizable as a low-altitude manned flying platform passes by the plant (Image: GREAL – European University of Rome)

Fig. 1.3 Screenshot from Google Earth showing part of the urban texture in the immediate surroundings of St. Peter's Basilica in Rome, Italy. The clearly recognizable dome of the basilica (*bottom left*) was the main landmark of Rome when the city was the capital of the Pontifical State (until 1870). The dome symbolized the spiritual mandate and power of the Church, made visible to all citizens. The large residential blocks north of St. Peter's complex were built after Rome had become the capital of the Kingdom of Italy. The new government was, at all levels, hostile to the papacy. City authorities had therefore approved a building plan for which blocks and streets were designed to prevent as much as possible the visibility of papal symbols. This was achieved by means of tall buildings and divergent alignment of streets that effectively obstructed the view of the dome from the entire area (Rendina 2004). Streets connecting the blocks were named after renowned non-Catholic personalities or famous figures of pagan antiquity

Each geographical context can be understood either synchronically or diachronically. In any case, however, a proper definition of the acquisition timing is necessary. It is worth noting, however, that especially in anthropization studies, diachronical analyses prove very useful toward a better understanding of evolution mechanisms.

Many of the elements and processes summarized above can be detected by the use of small drones. However, can this be achieved by UAV surveying and reconnaissance only? Obviously not. UAVs are just new, powerful, additional tools put in the hands of geographers. They can effectively do some parts of the job, but a full-blown knowledge about *places* goes far beyond the ability of systems to acquire data about *spaces*. Nevertheless, UAVs can provide researchers and operators with a large amount of information. Hence better analysis and interpretation can follow wider, quicker, and cheaper (in all aspects) data acquisition.

Another element is deeply connected to the just mentioned fact, but does not necessarily coincide with it.

Compared with traditional airborne platforms, small flying drones (and related technologies) are shortening the operational chain needed in the workflow. A primary advantage is that the degree of professional specialization required for the different tasks is decreasing as innovation progresses.

This brings two important consequences: (1) acquisition, processing, and presentation of data become more *accessible* and (2) all of the above can be more directly managed by the user who is supposed to produce and convey the final knowledge output.

Although the concept of personal mapping is a common assumption today as mobile applications are widespread, the actual development of personal airborne observation capabilities was not obvious and is presently rising (Mayr 2011; Gulch 2012).

1.2 Scales, Views, and Boundaries

A most basic concept in geography is scale. Environmental and anthropic phenomena are related in many ways: their interactions occur within individual locations, around small areas or settlements, up to wider regional levels, and even, in principle, on a global scale. Data, as fundamental units of geographic information, are to be acquired at the appropriate scale, for phenomena to be adequately understood and described.

Geographers conceived and used the "view from above" as a method, long before human flight was achieved. Reflections and conceptualizations are often conducted by "increasing" or "decreasing" the observation "heights." Each "height" corresponds to a level of information, and this, in turn, indicates levels for interpretation.

Before drones, geographers could typically acquire data by land surveys and airborne and orbital remote sensing.

Obviously, the final representation scale was not necessarily the one which had proved the most suitable in the acquisition phase. For instance, ground-based, wide-area cartography has always been the combined result of long, patient, and often painstaking local surveys. Maps at lower scales are developed based on larger-scale works, etc.

Flying platforms are able to give information about phenomena on the surface of the Earth by different kinds of remote sensing techniques. In general, though, the work they can do is dependent on their capability to discriminate objects on the ground and to measure some of their qualities. This typically involves basic spatial and radiometric resolution of sensors. Temporal resolution, i.e., the possibility of repeating the observation at defined time intervals, plays an important role too.

A factor toward a successful observation is often the flexibility that sensing platforms show in "zooming in" to or "zooming out" from a specific target. This indeed translates into a more detailed or more general view of the "scene" of interest (Fig. 1.4).

In the pre-drone framework, there is a surface data acquisition, with the geographer directly inspecting phenomena on the ground, an airborne observation which allows acquisitions at different regional scales; then there is a spaceborne segment which acquires geographic phenomena at smaller, i.e., wider, scales from regional to continental and global.

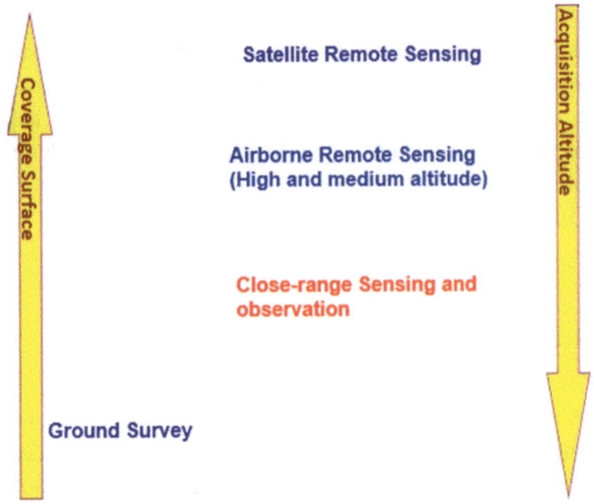

Fig. 1.4 Different observation systems could be "stacked" in relation to their typical acquisition altitude and coverage area. Small drones operate at the "close-range sensing and observation" level

Since drones appeared as a technology, a wide range of possible implementations became available. This suggests to conceptually connect different types of machines to different scales and themes.

Taxonomies were developed in recent years in order to properly categorize UAVs by technical and operational nature in more or less differentiated fields (Austin 2010; Watts et al. 2012; Salamí et al. 2014).

The rise of small drones created a new segment for airborne remote sensing and geographic observation: the so-called dronosphere (Germen 2016), spatially between manned airborne activity levels and the ground surface.

The term dronosphere, although wider in its general meaning, could be adapted to define the portion of airspace where small drones can operate. Its boundaries are currently dependent on technical features and regulations, but they generally include close proximity to the ground. At the moment there is no globally standardized concept for it, nor a shared definition for its limits. Yet, a fact is certain: the dronosphere is for relatively small drones *only*. Large drones, similar to manned aircraft in terms of size or flight rules, could not easily operate in that environment. Even if they could, it is likely that they would be prevented from accessing it, or they would be allowed into it with more or less severe restrictions.

No matter how the dronosphere will be established in the future, it seems that it will include portions of airspace which were not traditionally used in geographic observation, and this is *per se* an innovative element.

As of today, costs for the platforms are dropping, while usability is increasing at an impressive rate. The overall portability and integration of cooperative technologies may lead to a new paradigm in the production of spatial information: bottom-up and crowdsourced remote sensing (Meier 2014).

A larger number of operators may be enabled to acquire data and use them for producing and providing geographic information. This would be possible in the most diverse fields, from traditional themes to neo-geographic ones: what we can foresee goes much farther than present-day opportunities.

Getting back to scales, airborne observation is not necessarily bound to a specific kind of platform. UAVs and manned aircraft may often coexist in the same airspace at different (and not obviously layered) altitudes and environments. However, for the scope and purposes of this book, a simpler stratification of devices and techniques can be indicated. As far as small drones are concerned, their use will be preferable over local areas of interest at low altitudes and modest speeds; at higher altitudes, manned aircraft (or larger drones) will be able to cover increasingly wide zones.

In theory, data analysis and interpretation could seamlessly follow the same procedural steps regardless of the acquisition platform. The final user, then, would have little or no interest at all in the machine which materially carried the sensors.

This seamless integration is probably in the future, although the present is still dealing with several issues that remain to be solved. At any given scale, data processing is generally system dependent; its inner mechanisms may vary dramatically.

When discussing scales and themes, it is logical to connect UAV applications to geographic information science. Indeed, theme-related segmentation and deconstruction logics are obviously similar in both cases, and so are the general criteria about scales. In wide area observation, typically performed by the use of satellites or high-altitude airborne reconnaissance, a classical approach is to evaluate phenomena with main focus on their overall nature. Local area elements or point local details are considered nonrelevant and typically neglected.

By adopting progressively larger scales, we admit first larger regions, then smaller areas, then groups of objects, and finally individual objects which deserve special attention in our work. These different levels of "thematic resolution" imply different objects of interest and, therefore, different methodologies for the analysis.

At very large scales, furthermore, nadiral views – typically dominant in traditional remote sensing – may prove insufficient for some types of observation; the tridimensional nature of some specific objects of interest may require the adoption of views from different angles.

Today, as it was in the past, nadiral views are essential in metric measurements and georeferencing. Oblique aerial views, on the other hand, were historically considered more suitable for a qualitative understanding (Haggett 2001, p. 696–697) of some geographic phenomena, among them, in particular, anthopization-related phenomena. For decades, the two kinds of view were felt as complementary but necessarily separate. However, the rise of computer-based cartography and more powerful data processing techniques revolutionized this traditional habit. If sensor quality is sufficient and flight techniques are appropriate, it is now possible to obtain decent cartographic rendering of geographic features which had been surveyed by sequences of oblique views *only*. This is just to say that the traditional perceived "ontological" separation between oblique and nadiral views and between the respective purposes can now be considered, at least in some degree, obsolete.

Another aspect to be taken into account is that, for the purpose of qualitative geographical studies, precise georeferencing is not always necessary. When this is the case, simple geolocalization can be adequate for the goals of the work, especially considering that present-day GIS can easily handle hypertextual and multimedia data.

When discussing the use of small UAVs in geography and geography-related fields, observation scales must be conceptually connected to observation *boundaries*. Boundaries are to be intended as practical spatial limits to drone operation.

In approaching small UAVs, a geographer must be aware of some limitations which define and – so to speak – circumscribe the applicability of this technology.

First of all, technical range and flight envelope. Present-day small drones are often affected by modest area-coverage capabilities; this can be due to technical or juridical constraints. Examples of the latter can be found in some national regulations, like in the case of the typical mandatory VLOS[1] requirement or the prohibition to fly more than one UAV at a time. These two rules, for instance, would pose some difficulties to sending the UAV far from the pilot and to launch self-coordinating swarms of drones to increase the covered area.

Since these operational boundaries determine, in fact, where the observation can and cannot take place, all factors must be appropriately evaluated in preparing fieldwork.

Limited range, low heights, and slow speeds mean that a small drone can be a powerful data collection system at very large scales on local areas (Pérez et al. 2015); but it will become less and less competitive against other flying platforms as the scale is reduced and the observation area increases. This is to say that there are some types of remote sensing for which small drones are not suitable, at least today. Some examples will make the point clear. Let us assume that our purpose is to observe an ancient landmark – say, a small castle – and to obtain a high-resolution, close-range 3D model of the building. A small multirotor drone may take a while for doing the job, but it is able to fly very close to the facades, roofs, and architectural elements. Although sensitive to wind and turbulence, multirotors are fairly stable machines, and if weather conditions are sufficiently fair, a decent level of precision control can be kept at all times.

In sufficiently advanced machines, trajectories can be thoroughly planned and autopilots are quick and effective both in navigating the drone and in assisting a human pilot. Although, in principle, a manned helicopter could carry very high-resolution sensors, it would be extremely expensive and dangerous to deploy one over the castle for carrying out that particular mission.

Any fixed-wing airborne platform (either UAV or manned) would be less effective than a rotary wing when flying around such a point-based geometry; still, a small fixed-wing UAV would be slower and more maneuverable around the castle's features, while a larger drone or a manned airplane could only fly at a safe altitude along regular trajectories.

[1] Visual line of sight (VLOS) is a type of flight conduct in which a UAV must be kept constantly in such a position so as to be seen by the pilot. VLOS is obviously affected by technical nature of the drone, but regulations in several countries establish limits to never exceed both in terms of horizontal and vertical distance.

Another type of observation might totally change the perspective.

If the purpose is now to survey a 200 km (~124 mi) long, wide, and regularly anthropized coastline, then a small drone with currently standard range, especially if constrained within distance-limited regulatory requirements, will take a few days for covering the entire area. This would force the pilot, and possibly other ground crew, to remain on-site for the entire period. A commercial off-the-shelf multirotor can typically fly 15–30 min at speeds around 40–60 km/h (~25–37 Mph). An equivalent-class commercial fixed-wing UAV can fly up to 2 h at a speed between 40 and 80 km/h (~25–50 Mph). From a strictly technical point of view, then, it would perform much better in terms of area coverage.

Still speaking from a technical point of view only, if the problem is to cover wider areas in less time, a possible response would be to deploy swarms of UAVs, capable of flying respective missions in more or less coordinate manners.

Naturally, the advantage in endurance could be best exploited if the drones were allowed to fly long paths on autopilot, far beyond the pilot's line of sight. Once again, this is technically easy to achieve today, but it is often (appropriately) forbidden by authorities. The problem could be possibly overcome by organizing a "chain" of control stations and pilots, handing over the control to each other, along the survey line, or by implementing reliable and aviation-certified automated collision avoidance systems which are currently being designed and developed. It would also be necessary to reach some integration with air traffic control and services. In fact, all of the above solutions appear to be either theoretical or academical, at the moment. Normally available low-cost small drones do not support such capabilities.

In practice, it is difficult to believe that a profile like the proposed one could be considered acceptable.

A small UAV with limited technical reliability, blindly navigating a preprogrammed route over inhabited areas and infrastructures, with no ability whatsoever to see and avoid an incoming aircraft (which would hardly see a tiny and slow flying object), would be morally wrong before being – as it is often the case – totally illegal.

As far as range and speed are concerned, small manned aircraft (Cessnas, Pipers, etc.) might prove more efficient in the intended survey of the littoral.

It is worth noting, at this point, how the larger payload that a manned aircraft can lift in comparison with a drone would also allow to use larger and better sensors.

In conclusion, the use of a current low-cost UAV for the proposed survey of the littoral would become far less attractive.

As it will be discussed further in this book, an evident fact is that drones are technically able to perform much more kinds of operations than regulatory systems are generally willing to concede. Regulations in different countries affect how drones can be used (i.e., what kind of operation they can perform), their permissible geographical environments (city centers, peri-urban, countryside, etc.) and, quite logically, the airspaces they are allowed into. We are stressing this point because geographers should always take into the proper account that regulatory boundaries have a direct impact on the locations and kinds of phenomena which can be studied by using UAVs.

An important memento is that observation of anthropization-related phenomena would require presence and action in spaces which are often prohibited to drones or where access is heavily limited. If technology is capable of fostering knowledge about specific themes in specific locations, this does not necessarily mean that it will be normally allowed to go there. Yet, regulations and authorities do contemplate exceptions or conditioned authorizations if appropriate planning and implementation criteria are followed.

Last, but not least, relevant boundaries to the use of drones are determined by the socially shared perception of the drone capability to serve as knowledge acquisition tool. This aspect is rising as drones are turning from the status of charming and innovative technology for specialists into a widespread product in the hands of everyone. In this scenario, potential misconduct may derive from the availability of often impressive observation technologies; and even in the case of benign applications, it is now clear that small UAVs can "see" and "do" much more than other data acquisition tools.

This actually goes beyond the simple concept of "privacy" in terms of the right to personal seclusion. It may also involve social and political views about what people are willing to accept in someone else's (including law enforcement's and governments') capability to observe them with little or no corresponding need of being observed (Finn and Wright 2012; Bracken-Roche 2016; Hilton and Shaw 2016; Wang et al. 2016).

An article by Chris Sandbrook (2015) considers the social implications of using drones with specific regard to biodiversity conservation. However, his remarks are valuable in a more general discussion, as they point out strengths and weaknesses of the technology. On the one hand, UAVs are viewed as affordable, safe, and powerful tools; on the other hand, their compatibility with regulations and their overall perception are not obvious. A negative social perception could severely affect the use of these tools and should be carefully avoided.

As small drones are becoming a common presence in everyday life, a deep reflection on social and political implications is increasingly urgent. A review on recent work in this field can be found in Klauser and Pedrozo (2015).

1.3 Eyes in the Sky and Artificial Brains?

Small UAVs are bringing about a revolution in remote sensing. The concept is twofold. It integrates the reflection between *detection* of geographical elements with the problem of establishing paradigms, criteria, and logics for an *understanding* of their nature.

As it will become evident in Chap. 3, much effort of current research on UAV-based remote sensing seems to be oriented toward automated recognition of elements and objects within a specific theme.

Small flying drones would then be understood as carriers for different possible, increasingly powerful "eyes" in the sky. A well-known example of the importance of this evolution can be seen in the introduction of automated photogrammetry and computer vision in drone surveys. Some basic, general principles of vision and measurement are decades if not centuries old. However, the technical procedures which were necessary for obtaining the results were so complex, expensive, and cumbersome in those days, that in many cases available methods were not even used because they were just not worth the effort. Today, the advent of structure-from-motion (SfM) algorithms and more powerful computer graphics made it possible to turn once nearly prohibitive processes into feasible practice. Moreover, results can be achieved by the involvement of moderately skilled personnel – no longer high-profile specialists.

Another step is the increasingly effective classification of geographical objects as detected by UAV-acquired datasets.

In different types of remote sensing, there is a developing interest toward object-based image analysis (OBIA) and toward its specific geographical variant, i.e., geographic OBIA or GEOBIA (Blaschke et al. 2014).

It is intriguing to think that, as scientific fields and related technologies progress, the increasing effectiveness of the "eyes in the sky" could be accompanied by new powerful potential of "artificial brains". If unmanned aerial systems become able to automatically perform low-level data organization, then the work of human analysts could move toward higher interpretive levels and a deeper comprehension of geographic phenomena.

In fact, a long way is still to go. For now, automated analysis and interpretation are in their early days, although semantic capabilities of systems are developing. Today and in the foreseeable future, human mind and its higher conceptualization are still necessary for structuring the complex ontologies needed for machine-based interpretation.

References

Austin R (2010) Unmanned aircraft systems: UAVS design, development and deployment. Wiley, Hoboken

Bignante E (2010) The use of photo-elicitation in field research. EchoGéo [online] 11. doi: 10.4000/echogeo.11622 . Available via http://echogeo.revues.org/11622;DOI:10.4000/echogeo.11622 . Accessed 3 Mar 2017

Blaschke T, Hay GJ et al (2014) Geographic object-based image analysis – towards a new paradigm. ISPRS J Photogramm Remote Sens 87:180–191. doi:10.1016/j.isprsjprs.2013.09.014

Bracken-Roche C (2016) Domestic drones: the politics of verticality and the surveillance industrial complex. Geogr Helv 71:167–172. doi:10.5194/gh-71-167-2016

Buttimer A (1993) Geography and the human spirit. The Johns Hopkins University Press, Baltimore

Finn RL, Wright D (2012) Unmanned aircraft systems: surveillance, ethics and privacy in civil applications. Comput Law Secur Rev 28(2):184–194. doi:10.1016/j.clsr.2012.01.005

Frémont A (1976) La Région, Espace Vécu. Présses Universitaires de France, Paris

Germen M (2016) Alternative cityscape visualisation: drone shooting as a new dimension in urban photography. Electronic Visualisation and the Arts (EVA 2016), London., 12-14 July 2016, pp 150–157. doi:10.14236/ewic/EVA2016.31

Grainger A (2017) Citizen observatories and the new earth observation science. Remote Sens 9(2):153. doi:10.3390/rs9020153

Guarrasi V (1992) Cultural geography and semiotics of culture. In: Corna Pellegrini G (ed) Humanistic and behavioural geography in Italy. Pacini, Pisa, pp 29–35
Gülch E (2012) Photogrammetric measurements in fixed wing UAV imagery. Int Arch Photogramm, Remote Sens Spat Inf Sci XXXIX-B1:381–386
Haggett P (2001) Geography: a global synthesis. Pearson Education, Harlow
Hilton RK, Shaw Jr JM (2016) I see you. Drones, Privacy Torts and Insurance Claims. Claims Magazine, December: 21–23
Klauser F, Pedrozo S (2015) Power and space in the drone age: a literature review and politico-geographical research agenda. Geogr Helv 70:285–293. doi:10.5194/gh-70-285-2015
Mayr W (2011) UAV-mapping – a user report. Int Arch Photogramm, Remote Sens Spat Inf Sci XXXVIII-1(C22):277–282. doi:10.5194/isprsarchives-XXXVIII-1-C22-277-2011
Meier P (2014) Seeking digital volunteers to search & protect Namibia's wildlife (Using Aerial Imagery from UAVs). Explorers J. September 15. Available via http://voices.nationalgeographic.com/2014/09/15/using-uavs-to-crowdsource-the-search-for-namibias-wildlife/. Accessed 3 Mar 2017
Pérez M, Agüera F, Carvajal F (2015) Low cost surveying using an unmanned aerial vehicle. Int Arch Photogramm, Remote Sens Spat Inf Sci XL-1(W2):311–315. doi:10.5194/isprsarchives-XL-1-W2-311-2013
Raffestin C (1977) Paysage et Territorialité. Cah Géogr Quebec 21(53–34):123–134. doi:10.7202/021360ar
Raffestin C (1982) Remarques sur les Notions d'Espace, de Territoire et de Territorialité. Espace Sociétés 41:167–171
Rendina C (2004) Le strade di Roma, vol 1. Newton Compton, Roma
Salamí E, Barrado C, Pastor E (2014) UAV flight experiments applied to the remote sensing of vegetated areas. Remote Sens (Basel) 6(11):11051–11081. doi:org/10.3390/rs61111051
Sandbrook C (2015) The social implications of using drones for biodiversity conservation. Ambio 44(4):636–647. doi:10.1007/s13280-015-0714-0
Tuan YF (1977) Space and place: the perspective of experience. University of Minnesota Press, Minneapolis
Turco A (1988) Verso una teoria geografica della complessità. Unicopli, Milano
Turco A (2002) Paesaggio: pratiche, linguaggi, mondi. In: Turco A (ed) Paesaggio: pratiche, linguaggi, mondi. Diabasis, Reggio Emilia, p 7–52
Turco A (2010) Configurazioni della territorialità. Franco Angeli, Milano
Vallega A (2003) Geografia culturale. Luoghi, spazi, simboli. UTET, Torino
Vallega A (2004) Le grammatiche della geografia. Patron, Bologna
Wang Y, Xia H et al (2016) Flying eyes and hidden controllers: a qualitative study of people's privacy perceptions of civilian drones in the US. Proc Priv Enhancing Technol 3:172–190. doi:10.1515/popets-2016-0022
Watts AC, Ambrosia VG, Hinkley EA (2012) Unmanned aircraft systems in remote sensing and scientific research: classification and considerations of use. Remote Sens 4(6):1671–1692. doi:10.3390/rs4061671

Chapter 2
Concepts and Issues

Gianluca Casagrande and Davide Del Gusto

Abstract This chapter is intended to describe the main concepts and issues of applying small UAVs to geographic observation. The discussion begins with an overview of drones as flying platforms (2.1). We then introduce a fundamental concept, *accessibility*, or the set of criteria to *have* the needed technology (2.2). The following point, called *field deployability*, considers aspects regarding how the drone can get to *be used* in practice, based on technical and organizational steps (2.3). General but crucial concepts of the emerging national and international regulations about small drones are then discussed (2.4). In the last Sect. (2.5), a case study is presented in order to explain how mission planning should be approached based on available technology and operational needs.

Keywords sUAV • RPAS • Accessibility • Field deployability • UAV regulation

2.1 Friend or Foe: The Small Flying Drone

Gianluca Casagrande

Conceptually, the idea of unmanned aerial vehicles, i.e., flying machines capable of operating without an onboard crew, is quite old. Many activities, either in war or peace, may be so "dull, dirty, or dangerous" that one can find it more appropriate not to involve human operators in them. Current implementations are able to perform several different tasks including observation, measurement, and transport (Beloev 2016; Hristov et al. 2016).

The currently widespread UAVs derive from military ancestors (Birtchnell and Gibson 2015). However, as it has been the case with other world-renown technologies (e.g., GPS, GLONASS, GIS, etc.), the widest developments are often the result of booming civilian applications, invariably rising as soon as technologies become affordable. From that moment on, the development continues, often way beyond the initial pioneers' intentions and imagination.

G. Casagrande (✉) • D. Del Gusto
Geographic Research and Application Laboratory, European University of Rome, Rome, Italy
e-mail: gianluca.casagrande@unier.it

G. Casagrande et al. (eds.), *Small Flying Drones: Applications for Geographic Observation*, DOI 10.1007/978-3-319-66577-1_2

Complex taxonomies have been developed, through time, for drone categories, architectures, and application scenarios. Watts et al. (2012) include three categories for long-endurance UAVs: HALE, MALE, and LALE (where LE stands for Long-Endurance and HA, MA, and LA for High Altitude, Medium Altitude, and Low Altitude, respectively); two categories feature short endurance (LASE, LASE Close). Then come two minor categories (MAV/NAV), plus a VTOL category (Vertical Takeoff and Landing). This taxonomical approach suggests a conceptual prevalence of fixed-wing machines, a clear emphasis on large and medium drones and on their operational similarity with manned aircraft. As of this writing, this approach appears to be prevailingly oriented to military, institutional, or, in general, high-profile technical operations. In many instances of these scenarios, fixed-wing architectures are more applicable than rotary wings or other formulas (trikes, blimps, balloons, etc.).

If we were to adopt those authors' terminology, our book discusses prevailingly MAV/NAV and corresponding-size VTOLs. These, in fact, appear to be the most widespread kinds of UAVs on the civilian market. The term "small" applied to drones is quite generic, and a lot of remarks should be added for the sake of completeness. For the sake of simplicity, nevertheless, let us refer to UAVs within a classical weight limit of 25 kg. This should be considered – with some exceptions – as an upper limit in our discourse (Figs. 2.1–2.9).

Fig. 2.1 A typical small fixed-wing UAV for scientific and professional applications: the FlyGeo 24 Mpx has a cruise speed of 39 km/h (~24 Mph), an endurance of approximately 30 min, and a total weight of 2.8 kg (6.17 lb). Platforms of this kind can transport a relatively wide range of remote sensors, including visible, near-infrared, thermal, and multispectral devices (Image: FlyTop www.flytop.it)

Fig. 2.2 Medium-sized multirotors (above, a 6 kg or 13 lb FlyNovex hexacopter) are suitable for several types of applications spanning from precision farming to detail 3D modeling, to high-profile aerial photo/video documentation. Due to their efficiency and flexibility, they enjoy widespread popularity among professional users, in spite of often severe regulation constraints to their use (Image: FlyTop www.flytop.it)

Fig. 2.3 Depending on their size and takeoff speed requirements, fixed-wing UAVs can be launched by hand or by catapults. Drones of this type can land on small spots, either manually or automatically. *Below*, a FlySecur UAV standing on its field launcher (Image: FlyTop www.flytop.it)

Fig. 2.4 A small helium-filled drone airship (1.5m³ of capacity) can remain aloft and active for hours. This type of drone is suitable for tethered and free flights for general observation, but it can be precisely controlled only under particularly calm meteorological conditions. Larger and more powerful blimps have better performance in outdoor applications, but weather (particularly wind) remains a major variable (Image: Associazione Dirigibili Archimede, http://dirigibili-archimede.it/)

Having established this basic boundary, small UAVs can then be described as belonging to almost any possible architecture and as flyable through manual, semi-automated, or fully automated procedures.

In this book, we will mostly refer to some specific and widespread families of UAVs, but it is worth considering this very complex set of technologies in its broader range at least here.

A very basic distinction is made between rotary wing and fixed-wing UAVs. Rotary wing UAVs are normally helicopters and multirotors, relatively similar in the fact that both develop their lift by spinning propellers. A major difference, how-

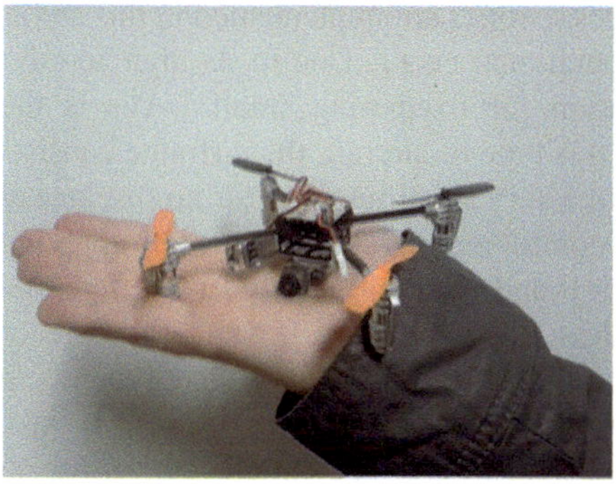

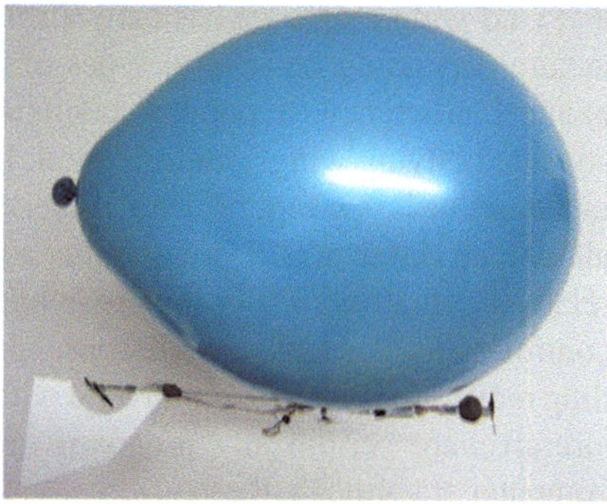

Figs. 2.5 and 2.6 In principle, drones can be very small machines, and yet they can cover specific tasks. *Left*: a modified Walkera Ladybird quadcopter fitted with an FPV camera. This UAV weighs about 70 g (0.15 lb), can be flown and controlled well in light wind, and is able to broadcast a 640×480 px analog TV signal up to 20–30 m distance (Image: GREAL – European University of Rome). *Right*, a micro-blimp (SpyBlimp) created by applying a small payload (including a tiny video camera) to an individual helium-filled balloon. This type of drone could be used in indoor spaces for inspection purposes (Design, build, and image by Yvon Masyn)

Fig. 2.7 Miniaturization is one of the main development factors for the civilian drone market. While technical performance is increasing, smaller drones are less invasive and easier to deploy. The FTD300X is a custom reengineered variant of the Parrot Bebop drone. Featuring a maximum size of 35 cm and a total weight of 297 g (0.65 lb), it belongs to a category of ultralight commercial micro-drones applicable to observation tasks from video footage to qualitative inspection and aero-photogrammetry (Image: FlyToDiscover)

ever, is that helicopters generally use rotors with variable pitch blades. Multirotors typically rely on differential power, accurately distributed among more simple fixed-pitch propellers. Multirotors may be less agile than helicopters, but they are generally as much precise and often configured so as to be more stable and easier to fly. This family of machines is quite differentiated and normally includes UAVs powered by 3, 4, 6, 8, or even more engines.

Fixed-wing UAVs are basically airplanes. Drones are implemented in this architecture depending on several types of operational needs. Due to its high aerodynamical efficiency, a popular configuration for fixed-wing small UAVs is the so-called "flying wing" configuration. In this type of aircraft, the airframe consists in a more or less swept or delta wing, with a very small fuselage (or no fuselage at all) and no classical tail; yaw stability and control is generally enhanced by applying relatively small vertical winglets, while roll and pitch are ensured by differential control of surfaces such as ailerons, spoilers, and flaps. The efficiency in terms of drag reduction is so high, compared with traditional configurations, that flying wing aircraft were periodically proposed in manned implementations as well. In pre-computer era, though, these machines proved unstable and potentially dangerous, so their use remained limited for a long time.

The problem appears to be completely overcome by drones – and they are generally expendable, after all.

Multirotors and flying wings are currently dominating the market in geographical and geography-related fields due to their flexibility and reliability as aerial observation platforms. Nevertheless, there are many other types of drones which may serve well in several fields. Classical airplane architectures are common for larger UAVs, including motorgliders and motorized parachutes (Lelong et al. 2008). The latter are also used due to good endurance and payload capability along with small encumbrance.

Peculiar types of UAVs are also used for specific purposes, e.g., flight-testing (an interesting example in Adams et al. 2013) or in-flight recognition and manipulation of objects. Other solutions include hybrid maneuvering capabilities, e.g., flying drones able of advanced ground motion, flying multirotor UAVs equally able to move underwater, etc.

Drone-blimps are a kind of so-called "lighter-than-air" UAVs. Compared to "heavier-than-air" machines, they share strengths and weaknesses of their manned counterparts. Airships are indeed capable of lifting relatively heavy payloads, and they are able to fly for long time, as their endurance relies on the lifting capability of gas rather than on the power developed by engines. On the negative side, drone-blimps, like the giant airships of the past, are extremely weather sensitive and require fairly complex organization and equipment for their deployment.

By observing the progressive release of new versions for models on the market, the impression is that drones evolve more as ICT platforms than aerotechnical products.

The efficacy of small UAVs in general – and in geographic applications in particular – relies upon the fact that they are generally flexible and reliable carriers of instrumental payloads. Such payloads can be visible-light cameras or other types of data acquisition and measuring devices.

Current small UAVs are remotely managed by a pilot/crew via a remote pilot station (RPS), also often referred to as ground control station (GCS), by a more or less reliable flight control links. Drones are generally capable of recording data onboard and they also exchange both flight and payload data with the RPS in real-time. Normally, the pilot can see (on screen or through First Person View – a.k.a. FPV – devices) aerial views from the drone. Additional data – commonly height, speed and position, among others – are also displayed.

Depending on types, small UAVs can easily be managed by a pilot, either alone or with the help of a system operator and one or more observers. Although regulatory and practical factors often suggest or require that the pilot keeps visual contact with the UAV, a very common feature of UAS is real-time telemetry and imagery downlink. These allow for the so-called First-person-view (FPV) operation: the crew may observe the target by the view acquired from one or more drone video camera(s) while, at the same time, monitoring the flight parameters and conditions of the drone.

UAVs flourished as electronic flight control systems (FCS) evolved from initially full manual to increasingly automated (Hristov et al. 2016).

A PwC report (PwC 2016, p. 4) summarizes the worldwide market perspectives of small UAVs. Estimations are based on the likely commercial value of business and labor which can be replaced, in each one of the involved industries, by the use of drones. As of 2015, the figure was 127.3 billion USD, so divided: infrastructure (45.2), agriculture (32.4), transport (13.0), security (10.5), media and entertainment (8.8), insurance (6.8), telecommunication (6.3), and mining (4.3). This points to a clear trend, especially considering that some applications for future drones have not been identified yet. Nevertheless, it should be noted that even when a potential use is foreseeable, that does not necessarily mean that it will come soon or ever. For some applications, current drones are still technologically immature; for some others, their use will not be allowed for a relatively long time. For others, finally, they might just not turn out useful or competitive in practical terms.

News about the impressive performance of small drones in remote sensing may be deceiving to some potential users. As a matter of fact, given a certain goal, the specific available UAV can be effective or not. In our experience, observation by drones can prove partially conclusive or inconclusive depending on inadequate system configurations or insufficient planning. The design and technical features of a drone can prove suitable or not for the requested tasks; the deployment scenario may impose limitations from an organizational point of view; juridical requirements for using that particular drone in that particular site are an independent and often tricky variable.

In order to avoid problems, it is convenient to devote preventive attention to all the possible caveats other than the strictly technical ones. Operational pretesting of available equipment and protocols is also good practice, to ensure that the UAS operates predictably and reliably.

2.2 Accessibility

Gianluca Casagrande

Accessibility is a key concept in UAV-based geographic observation. A wide set of variables should be considered. The aim is to achieve efficacy at the lowest possible cost, where "cost" means the overall effort – whether financial, technical, or bureaucratical – needed to set up the systems, deploy them on the field, and successfully use them.

The accessibility of a technology determines how many operators and workgroups will be able to use it. A "linear" increase in the number of users may cause – and it often does – a "nonlinear" increase in the number of applications and developments.

The most basic meaning of the word "accessibility" is normally associated with the financial side of the activities. If it is not economically expensive, then more workgroups and individuals can afford to acquire the technology. Hence they can now do things that were previously possible to high-end users and high-tech systems only (Aden et al. 2014). This also implies the ability of widening and enhancing the networks of users. Workgroups and individuals can now use better tools for their scientific and application work, leading to a positive chain reaction in terms of innovation (Carrivick et al. 2013).

With regard to UAVs, a large set of examples could be indicated. In cultural heritage research, for instance, it is no surprise that cutting-edge technology is employed to survey world-renown archaeological sites. It is surprising, though, to verify that more modest technologies can provide very good quality data about minor landmarks. By spreading the use of these second-class tools on second-class targets, a large patrimony of valuable assets can be investigated and monitored. Similarly, it is perfectly obvious that high-tech UAVs are used by civil protection services during emergencies and disaster recovery. It may look less obvious that low-tech UAVs could be used by low-budget local institutions to conduct preventive and early warning surveys in their territories. In general, this technology would allow to monitor structures and elements previously or possibly involved in emergencies (Bendea et al. 2008; Beloev 2016). Preventive measures and early warning actions could make a difference in discovering and interrupting a chain of events to disaster.

Furthermore, if it is true that, for highly specialized purposes, correspondingly high sophistication is required, it is also true that in many ordinary, real-life cases such sophistication is not actually needed or the qualitative gap can be considered acceptable (Casagrande et al. 2015; Pérez et al. 2015; Fryskowska et al. 2016).

Additionally, there are uses for which it may be more appropriate to deploy the simpler and more modest vehicle, rather than the top-class one.

The importance of purely basic, inspective, or descriptive data acquisition is often neglected as scientific research on sensors, methods, and techniques constantly pushes toward enhancements.

However, a narrow-minded excessive emphasis on the importance of systems and technologies might lead to the wrong attitude of neglecting the potentials of simpler and humbler solutions. Banal as it can seem in theory, practice shows that the widespread and meaningful use of simple techniques and technologies might prove more scientifically, economically, and socially relevant than isolated, spot applications of cutting-edge solutions. Someday, those very cutting-edge solutions would become standard honest tools, giving way to even more advanced systems; and the argument will still hold true. In the end, the rush for technological state of the art may deceive us into underestimating the other, crucial components of good science and applications: a smart and well-thought observation methodology, good quality of data acquisition and processing, reliable conclusions, and the most widespread possible use of them.

Accessibility is related to a wide set of technological variables in the choice of UAV systems.

The difference between off-the-shelf UAVs and custom-built ones seems to melt down to a simple conceptual distinction. Off-the-shelf technology provides the user with ready-to-fly, well-tuned-up systems. As a drawback, much of that technology is proprietary and allows limited customizability. This is no big deal to general users but may cause legitimate frustration among specialists.

Prototypes or custom implementations, often (not always) relying on open-market components, have the advantage that solutions can be fully tailored on the specific needs (Giges 2016). Potentially efficient, nonconventional solutions can also be tested by following this approach (Yun et al. 2013; Zhao et al. 2015).

On the other hand, a remarkable quantity of work is generally needed for setting up and fine-tuning the systems. Reliability testing is also needed to make sure that UAVs can actually be deployed in a real-life context and operate safely.

A topic that must be taken into account is that, in at least some geographic applications, drones should be used keeping in mind their primeval vocation. In the beginning, they were supposed to be sent doing dull, dirty, and dangerous tasks.

This means that, as long as safety, security, and privacy are guaranteed, an operator should be enabled (and prepared) to sacrifice the drone for the sake of the results if they are particularly relevant. If the overall costs (either financial or organizational) are perceived to be excessive, then the drone prevails on its purpose, and this might not always be in the best interest of the users or the community.

In our experience (Casagrande et al. 2015), accessibility corresponds to essentially three criteria in systems choice and operational framework:

1. Immediate availability
   (a) Use of general market products.
   (b) Use of open-source or de facto standard hardware and software (when possible).
   (c) Avoid restricted proprietary solutions unless fully compatible with de facto standards.
   (d) Keep configurations as simple as possible.
2. Portability
   (a) Adopt de facto standard solutions for data acquisition, transfer, processing, and presentation.
   (b) Design systems and workflows in such a way that outputs are reliably comparable with those of other techniques.
3. Extensibility
   (a) Design configurations with modularity in mind.
   (b) Keep systems and operational profiles, (if possible) open to change.

Accessibility is quite an adaptive concept. It basically translates into getting the best possible set of tools while, at the same time, keeping all the costs at a reasonable minimum. An important element is to properly select the flying platform and the associated sensors while taking into account the operational scenario and other contextual conditions. An example of the concept of practical accessibility as we just stated can be summarized as follows.

A geographic laboratory drone fleet includes three VTOL UAVs. The first one is a minimal size machine fitted with a built-in video camera; the second one is a medium-size device able to mount different types of sensors within – say – 100 g (0.22 lb) of mass. The third one is a fairly large, heavy-lift drone able to fly with 6 kg (13 lb) of net payload; this too can have different sensors installed. The proposed field activity is to have a first, very expeditive documentation of a poorly studied and very wide area. The field operation will last a few days, and very limited ground support will be available in the meanwhile. If this is the case, then the minimal drone would probably be the most appropriate choice. It is lightweight and fairly small in size; its requirements in terms of power are relatively modest, which is important if the available energy supply is limited, in order to maximize flight time. The most accessible solution in this case is to operate the smallest and less invasive drone available. Note: given the proposed operational scenario, a fixed-wing UAV would have probably been more efficient, but the available fleet includes multirotors only.

Based on the same assumptions about the fleet, let us consider another case.

The planned activity consists in the thermographic inspection of a group of inhabited buildings, right in the middle of a historical town. However, thermal cameras are not available at the lab: it is therefore necessary to purchase one. If the overall conditions of the area suggest (or impose) to fly a small, lightweight, and noninvasive drone, then the choice should be to deploy the medium-size machine: it is a good trade-off in terms of dimensions, and it can install a small thermal camera in order to perform the inspection. A typical disadvantage would be the relatively high cost of the sensor, but the medium-size platform is the best choice since it can be cleared to operate in that particular urban area. On the other hand, if the overall conditions of the survey area are considered to be sufficiently safe to operate the largest drone and there is no adverse legal requirement, then the heavy lifter could be deployed.

If this is possible, then a heavier thermal camera can be acquired: a less miniaturized one and therefore, probably, less expensive.

In this case, the accessibility element of the equipment is the larger thermal camera. The UAV is capable of carrying it, and the more affordable sensor leaves margins for other items.

These examples were given to show that the concept of accessibility is application and context dependent. Choices should be steered having feasibility and cost-effectiveness in mind.

2.3 Field Deployability

Gianluca Casagrande

2.3.1 Authorizations and Organizational Overhead

Field deployability is basically the operator's ease in taking the UAVs to where they are supposed to operate and in performing the intended work.

It depends both on regulation and technical aspects. From the point of view of regulation, two questions are particularly important. The first one is the juridical framework in which the drone is to be operated. This is frequently determined by the type of technology and by the taxonomical category to which the UAV belongs in the regulatory context.

The second question is the organizational overhead implied by the regulation in addition to the specific UAV requirements.

This particular issue involves the operators, the pilots, and the operation per se. A general convergence in countries which published UAV regulations is toward formal qualification for pilots and validation/authorization for the drones. Specific rules are being established in terms of where, when, and by whom drones can be flown (Ogden 2013; Villa et al. 2016). This topic will be further discussed in Sect. 2.4. A general emerging criterion is that the less potentially invasive the UAV is, the less restrictions will apply to it and the less constrained the organization will be in using it, the less qualified and specialized the technical crew will need to be in order to work. In an average real-life research or professional workgroup, anytime simplifications are achieved, the overall feasibility of the job improves exponentially.

At the moment, drones are subject to specific limitations preventing them from normally flying over some areas. Regrettably, some of these are densely anthropized locations, which could be, for this very reason, extremely interesting in geographic analyses. This kind of limitation might be eased in the future by the rise of smaller, more reliable flying platforms.

The point is whatever the goals of the intended observation are, a correct determination of permissible operations is a major part of the planning.

Technical qualification and authorization issues involve civil aviation authorities (CAAs). Privacy, another hot topic in drone operations, is regulated at different levels by national and international laws. Then comes, in principle, another set of bureaucratic obligations which may affect the planning and conduction of a geographic survey: formal authorization might be required by authorities/entities controlling or managing the study *areas*. This could mean multiple independent authorization processes, each one of them having go/no go value. All of these activities add to what we had previously called the "organizational overhead." A survey flight on a specific area may require participation and the presence of specialists, technical persons, and, if necessary, the activation of support services. The organization which is running the activity must be able to deal with all related actions *and so on*.

In the authors' experience on the field, simplification always proved an essential ingredient to success (Figs. 2.8 and 2.9).

Fig. 2.8 Typical professional drones can be deployed on the field by the use of individual vehicles which could often carry all of the needed equipment (Image: FlyToDiscover, www.flytodiscover.it)

Fig. 2.9 A Sensefly eBee flying wing just deployed on the field, ready for takeoff (Image: Giuseppe Gallo)

2.3.2 Safety and Reliability

Once the operation is planned and approved, another set of questions must be considered. This time the main variable is the drone itself. First of all, is the UAV inherently safe for people and things? Technology is generally progressing, and the overall reliability of drones is developing along. Blimps and balloons have normally a relatively low damage potential due to their natural buoyancy, relatively high drag, and small weight. Fixed-wing drones can be very fast and relatively heavy, and their airframes are sometimes able to cause damage; in principle, however, their capability to glide is a mitigating factor in the case of possible failures. In many cases, moreover, fixed wings are built with lightweight composite materials whose specific mass is modest. The combination of the two things makes those airframes able to quickly deform and brake up in case of impact, so as to minimize potential damage.

Rotary wing drones behave in different manners depending on their architecture: their relatively high weight and small drag would cause these types of drone to suddenly plunge in case of a mechanical failure.

In favorable conditions, helicopters can remain fairly controllable and be guided to emergency landings; this is not the case for multirotors, which rely on redundancy to sustain malfunctions. If multiple engine redundancy is not available, multirotors just fall if one engine fails.

In order to improve the safety of small drones, some industries have developed emergency parachutes and other devices – activated by independent control links – to provide a second line of recovery in case the drone should suffer catastrophic failures or become uncontrollable. Among many other elements, the availability of electrical brushless motors allowed to replace, in many applications, the use of traditional internal combustion engines. This turned out as an important step toward reliability. Flight control systems are the other important elements of UAV safety. First generation drones were able to give a minor contribution to their own controllability and navigation; current systems provide the pilots with sophisticated assisted controls and quite complex automated procedures: takeoff, landing, emergency return to homebase, programmable autonomous navigation, etc. The pilot does not directly interact with the controls of the drone, but, in most cases, it interacts with the flight control system. This means that in the case of a failure or interference to this crucial component, a drone could become uncontrollable and either crash immediately or – which is much worse – go astray and become an unpredictable cause of damage. Some currently produced drones are equipped with redundant avionics and better system protections. Still, the certified reliability of most small drones is lower than accepted aeronautical standards, and their actual potential to cause damage is highly dependent on their weight, material, and shape. If, on the one hand, a perfectly reliable drone is yet (if ever) to come, on the other hand, it is clear that the real safety level of a specific UAV depends on its architecture and flying principles but also on the quality of its components, the configuration of its systems, and even the techniques used in its construction. Maintenance, obviously, would be a major factor. The knowledge of these multiple elements with specific regard to the deployed drone can provide indications about the actual safety of a specific UAV and indicate criteria for its use.

2.3.3 *Flight Handling Qualities and Usability*

The banal consideration that drones are not created equal – at all – brings up an important aspect: what are the actual handling qualities and performance of the machine to be deployed? Drones, like manned aircraft, have specific flight qualities which may be more or less suitable toward specific applications. We already mentioned that rotary wings are stable and maneuverable platforms for small area surveying, whereas fixed wings are more aerodynamically efficient and generally provide better services for longer transects or wider area observations. Blimps would be able to give outstanding endurances and good lifting capabilities, but they are very sensitive to weather conditions, etc.

Within each category though, different models feature different behaviors; this may affect the usability of the drone in many ways. A most critical element is how much training, skill, and technical effort is actually needed. Problems may rise if the UAV is inherently difficult to control, especially in precision maneuvers. Some systems require complex and error-prone procedures to perform expected tasks; in some other cases, automatic functions are insufficiently precise or unreliable. These are only a few of many possible issues (Figs. 2.10, 2.11, 2.12, and 2.13).

The more a UAV is difficult to fly and operate, the more specialized and skilled the flight crew will have to be, thus increasing the organizational overhead and limiting the practical usability of the drone.

Then comes operational reliability, which does not necessarily involve safety. A reliable drone is actually able to perform its tasks in a predictable and stable manner. Reliability involves the overall behavior of the UAV during its operation and its inherent qualities as data acquisition platform. Functional issues may occur that could not be classified as technical failures but rather unsatisfactory behaviors. Such is the case of inadequate performance of some devices (sensors, data links, GNSS, etc.). Typically, behaviors of this kind can be found in customized implementations when due to UAV-payload compatibility problems, one of the two components is negatively affected by the other. For example, many multirotor operators are familiar with sudden unpredictable distorsions of acquired images. These may be caused by vibrations from the engines in particular flight conditions such as hovering in turbulence or fast descents in the prop wash vortex rings. The fact that a drone shows undesired behaviors may expose it to becoming an unreliable acquisition platform.

Another apparently obvious aspect of usability is how the UAV can withstand mishaps such as heavy landings and crashes. Drone survivability in case of crash is not secondary; the success of field operations is often dependent on the fact that the equipment is able to operate even in case of a crash. The fragility/robustness of a

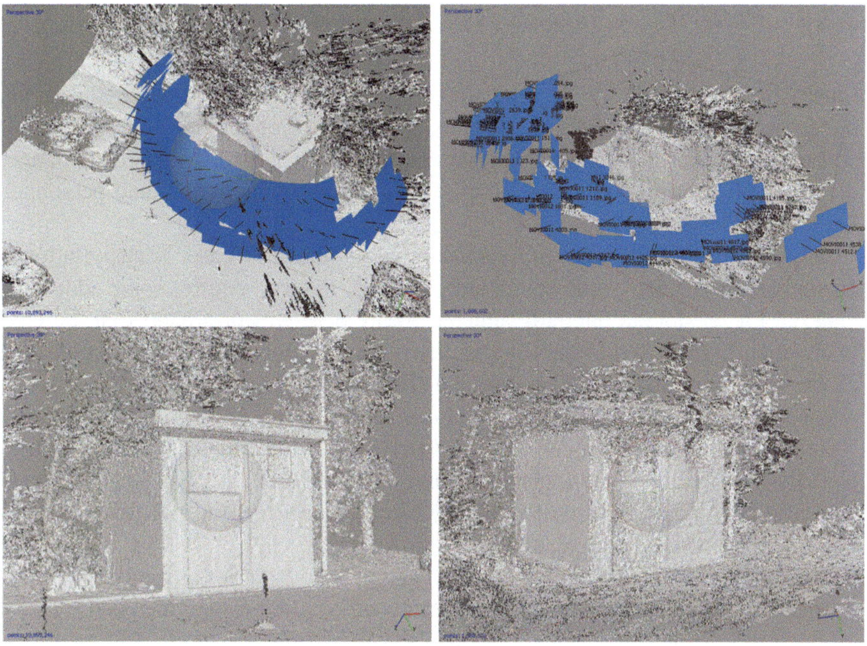

Figs. 2.10, 2.11, 2.12, and 2.13 Comparison of tridimensional models of the same sample building, obtained from images acquired by two different UAV systems. Images on the *left* (Figs. 2.10 and 2.12 camera positions and mesh, respectively) refer to the use of a medium-budget professional drone with assisted control and gyrostabilized full-HD camera; corresponding images on the *right* (Figs. 2.11 and 2.13) refer to a low-cost, fully manual drone with 720p non-stabilized camera. After flights, video frames were fed to a standard software (Agisoft PhotoScan), and a mesh was developed from both input set. The most important issue in the use of the low-end drone was the need of manually selecting sufficiently sharp images among a large number of poor-quality frames shot from more irregular trajectories. Such operation proved time-consuming and quite invasive on the workgroup's activity. Acquisition and processing of frames from the higher-quality system were more rapid and straightforward. Still, both methods led to the creation of a recognizable model of the target building. The actual value of each model depends on the final purpose: the choice between solutions must take into account the overall cost, deployability, and efficacy of the system (Images: GREAL – European University of Rome)

UAV system is obviously related to size and weight. In general, though, it is important to be able to rely on stubborn airframes and effectively resistant components, so that a banal mistake or marginal environmental conditions do not necessarily knock the drone out of the operation.

An important factor is the possibility of performing maintenance works on the field; for this to be achieved, the drone must be easily inspectable and appropriately modular so that components can be substituted without having to perform complex operations on other unrelated systems. Finally, the UAV should not need complex and cumbersome accessories to allow for its operation.

2.3.4 Batteries: A Specific Issue in Logistical Overhead

Finally, field deployability depends upon the logistical overhead required to bring the drone and the other service components on the area of activity.

Logistical overhead involves drone transportation issues, battery and battery management systems, type and complexity of support equipment, and personnel. Let us consider each aspect.

In general, small drones can be transported with relative ease. Transportation is dependent on the size of the equipment, and this, in turn, requires some additional material for its storage onboard vehicles. Many problems originate from the payload which is supposed to be flown in the survey. Payload, in fact, determines the characteristics of the needed drone and that determines what materials must be made available on the field so that the drone can fly.

In present-day situation, an element which must be taken into account very specifically when dealing with logistics involves how the drone is supposed to be powered. If the UAV has a common brushless electric powering, the question translates into batteries, battery storage, and recharging devices.

Currently available technology uses so-called LiPo batteries as prevailing power sources for small drones. Originally developed as lithium-polymer cells, actual products may show various materials and manufacturing standards. Although new implementations provide better handling features and safer construction, LiPo cells are famous for their bad behavior in case of inappropriate handling, wrong recharging, and even accidental impacts or damage. A possible consequence of these errors is that the batteries would start to operate unpredictably; another possibility is that they simply catch fire, with no need for any other ignition source.

This suggests the adoption of appropriate physical protections and safety procedures in recharging, storing, transporting, and loading-unloading batteries (Fig. 2.14).

Small drones have different power absorption rates depending on architectures, and this leads to significant differences in battery types, size, and features: blimps could stay aloft for a long time with minimal use of the battery; fixed wings are generally capable of remarkable endurances with medium-size batteries; rotary wings and particularly multirotors would need relatively large batteries to fly, and their power consumption is quite rapid. Adequate logistics should therefore be adopted in each case, and various trade-offs may be necessary.

2.4 Regulation: A Matter of Philosophy?

Davide Del Gusto

Regulation concerning professional and recreational use of drones is being developed, in different countries, based on mainly three primary concerns: safety, security, and privacy. Safety deals with ensuring that the UAV cannot cause damage to people or properties due to unintentional causes, such as malfunctions or pilot errors.

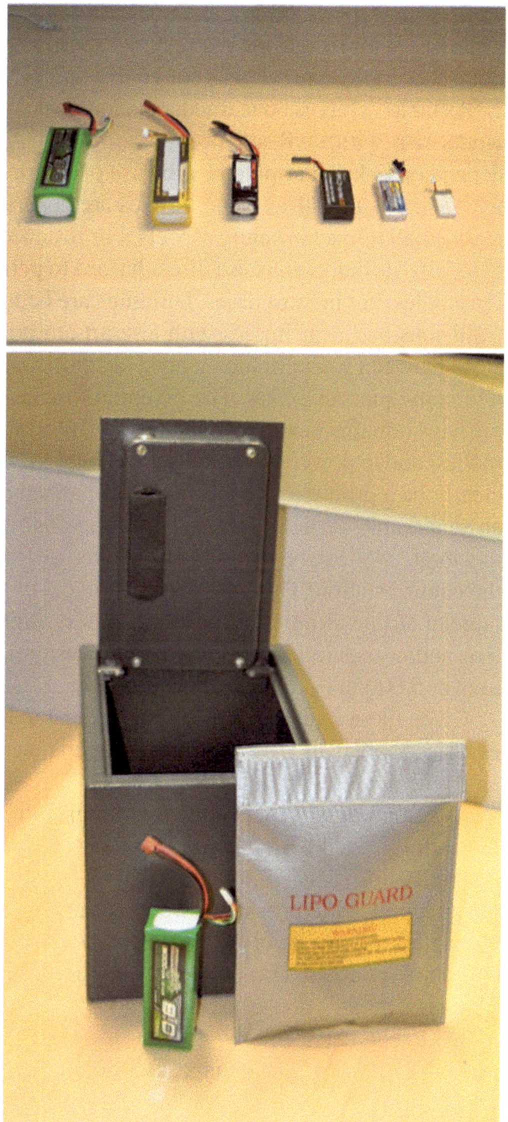

Fig. 2.14 *Above* is a collection of traditional LiPo batteries for small UAVs. *Left* to *right*: 4 cell 8,000 mAh; 4 cell 5,000 mAh; 3 cell 2,250 mAh; 3 cell 2,000 mAh, 2 cell 1,000 mAh, 1 cell 760 mAh. *Below*: a possible safe approach for storing and transporting large traditional LiPo batteries is to put them into a fire-proof bag and to put bags into a metal case (Image: GREAL – European University of Rome)

Security is involved in preventing the use of drones or some of their operational features (e.g., data links) for malicious actions such as terrorist attacks, criminal acts, or other types of violation. Privacy is becoming increasingly important as it is clear that new previously unknown juridical aspects emerge from the drones' capability of providing closeup aerial views of personal and/or secluded spaces. This involves quite complex matters, as data can be collected by private as well as governmental operators (Finn and Wright 2012; Smith 2013; Arteaga Botello 2016; Bracken-Roche 2016; Hilton and Shaw 2016; Wang et al. 2016).

Security aspects of drones in general (and small drones in particular) are being discussed with increasing interest by experts. The subject is, however, substantially out of our topic, and we will not go through it.

As far as safety is concerned, development and management of regulatory frameworks rest with civil aviation authorities (CAAs). In this field, obvious criteria depend on risks determined by the use of a certain drone and its potential to interact with general air traffic. Small drones are generally considered to have correspondingly small impact in safety aspects. For instance, their size, mass, and materials are generally sufficient to cause direct hazard to persons in outdoor spaces and possibly to vehicles, but in most cases, buildings are believed to provide adequate sheltering. Collisions and near misses with aircraft are possible and did occur in some occasions between low-altitude manned air traffic and small drones, with relatively limited consequences so far. The potential of causing major damage and catastrophic mishaps remains nevertheless very high. Overall range of small drones is relatively modest due to power consumption and total available energy supply, but altitude can clearly be a problem as many small unmanned aircraft can climb and be controlled at remarkable vertical distance from the control stations. A normal low-cost, off-the-shelf UAV can be theoretically piloted at 1,000–2,000 m (~3,300–6,600 ft) over the control station. These heights are typical of general aviation flights and of commercial airline climb-outs and approaches. Slight modifications to engines, propellers, and/or data links may enable small customized drones to operate at altitudes up to over 5,000 m (~16,000) above sea level.

Given these remarks, it is easy to understand why regulators are so concerned in ensuring that small UAVs are properly separated from sensitive ground objects and air traffic at all times.

Development goes toward an integration with manned air traffic systems depending on justifiable need, locations, and operational scenarios. This generally translates into establishing borders and boundaries that UAVs are never supposed to cross: allowed heights are variously limited depending on the characters of the involved airspace; no-fly zones are established over relevant and sensitive areas. These typically include airports, governmental infrastructures, industrial areas, densely populated places, areas occupied by crowds during public events, etc.

In spite of their limited range, as drones can fly over inhabited spaces, a major issue is to fine-tune regulation according to the ascertained levels of potential hazard of the UAV itself. Flyaways are righteously considered a major threat, in principle.

A standard commercial hexacopter can normally fly (and even climb!) at speeds around 40 km/h (~25 Mph), for 10–20 min; this means that an uncontrollable flying drone could pose a threat to people on the ground or in the air at quite far a distance from its takeoff base. In order to avoid this type of event, technology has developed – and regulation often requires – multiple preventive measures such as geofencing, terminating devices, tethering cables, restricted buffer zones, etc.

So far about general concepts, then comes real life.

By reviewing regulatory evolutions about UAVs in several countries, it turns out that some concepts seem to be rather uniformly perceived by the international communities; some others are not.

There is an overall consensus on the fact that recreational drones should be regulated differently from professional ones; it also appears widely recognized that regulation should hold safety aspects as a main driver and classify requirements based on that. The practical result, however, is a relatively wide range of solutions and trends. They all contribute to indicating how debate and conceptualization of drones as "real" aircraft in a "real" airspace are far from being over.

A complete and detailed study of national regulations would be out of the scope of this book. Our intention is just to present some general concepts which appear to be considered by several civil aviation authorities in different countries.

We have based the following comments on selected data drawn from www.droneregulations.info website and by information issued by national CAAs. In order to more effectively relate the contents of our summaries to the cited documents, we have adopted the following citation style: each cited document is indicated by the name of the country and year of publication in footnote, in alphabetical order. This allows to immediately view countries which issued regulation about a certain topic, relevant to our discussion. In the reference list at the end of the chapter, each document is indicated beginning with the name of the country and year of publication. All used documents are included in the reference list at the end of this chapter.

We have no ambition whatsoever to provide a complete and fully exhaustive review of the entire regulatory sources; our purpose is to briefly present some remarks on subjects which appear to be widespread in the regulations we have considered.

Since the effective use of small drones for geographic observation is possible only by taking regulations into account, users are obviously necessitated to include them in their planning process from the very beginning, as technical solutions might be conditioned by regulatory requirements.

2.4.1 Professional vs. Recreational Drones

Most of the examined regulations explicitly indicate a difference in juridical nature between drones used for professional and sport/recreational purposes[1]. Terminology itself varies, as drones used for fun and sport are variously called "models"[2] and in

[1] e.g. Argentina (2015), Australia (2017), Azerbaijan (2015), Bangladesh (2016), Belgium (2016), Brazil (2016), Canada (2014), China (2015), Croatia (2015), Cyprus (2015), Czech Republic (2013), Finland (2017), France (2017), India (2016), Italy (2016), Jamaica (2016), Mexico (2016), Netherlands (2015), Philippines (2015), Portugal (2016), South Africa (2015), Ukraine (2016), United Arab Emirates (n.a.), United Kingdom (2015), USA (2016).

[2] e.g. Australia (2017), Austria (2014), Italy (2016, 2017), United Kingdom (2015), USA (2016).

some cases "toys"[3], while professional drones are called UAVs or, more frequently, RPAS/RPA (remotely piloted aerial system/remotely piloted aircraft).

In fact, the "ontological" distinction is quite variable and not always associated with technical features. In general, all commercial applications are automatically classified as professional. In some cases, however, operations which involve any kind of nonsport-/leisure-related activity can be classified as professional, even if conducted for no revenue (e.g., Italy 2016; 2017). In general, there appears to be consensus that the main difference between professional and nonprofessional drones rests with their application; technical distinctions are also made in some national regulations, but the criteria are various and subject to change.

2.4.2 *Pilot and Operator Qualification*

Many – but not all national regulations tend to indicate complementary requirements for pilots and operators in spite of some interchange ability and variability in the formal naming for both[4]. Although often coincident, the two figures are conceptually different. In general, a drone operator, like with any other commercial aircraft, is the individual or company which is commercially operating the vehicle and manages its technical and revenue activity. The pilot is the qualified person who flies the drone on behalf of the operator and, according to the operator's intentions, under her/his professional responsibility as "pilot in command."

The qualifications and technical and bureaucratical requirements for both an operator and a pilot are variously defined by national CAAs based on different criteria, involving the specific drone, the size and purposes of the drone fleet, the operational scenario, etc.

Depending on national legislations, professional UAV pilots are subject to passing different kinds of medical examinations[5], different types of theoretical/regulatory training, and practical training sessions[6]. For example, in some countries, pilot's physical fitness must be ascertained by aeromedical tests equivalent or coincident with those required for manned aircraft (PPL, LAPL, microlight license, etc.).

---

[3] e.g. Austria (2014), Portugal (2016), Rwanda (2015), Slovakia (2015).

[4] e.g. Argentina (2015), Azerbaijan (2015), Botswana (2015), Brazil (2016), Cameroon (2016), Canada (2014), Chad (2014), Colombia (2015), Côte d'Ivoire (2017), Gabon (2014), Ghana (2016), India (2016), Italy (2016), Jamaica (2016), Malaysia (2008), Mauritania (2014), Mexico (2016), Netherlands (2015), Nigeria (2015), Portugal (2016), Rwanda (2015), South Africa (2015), Togo (2015), United Kingdom (2015), USA (2016).

[5] e.g. Argentina (2015), Belgium (2016), Canada (2014), Cyprus (2015), Ghana (2016), Italy (2016), Mexico (2016), Nigeria (2015), Rwanda (2015), South Africa (2015).

[6] e.g. Cameroon (2016), Dominican Republic (2015), Italy (2016), Philippines (2015), United Kingdom (2015), USA (2016).

As far as aeronautical competence is concerned, small UAV pilots are expected to have a basic but solid knowledge of the rules of the air as per national specifications. This kind of knowledge is typically limited for operations in noncontrolled airspace, but it can also include other types of airspace. Since practically all national regulations mandate certain separation of any small UAV from airports, their ATZ, and other restricted areas, a sufficient ability to read aeronautical cartography and documentation such as NOTAMs is assumed. In some cases, even small UAVs may need to apply for authorizations to enter airspaces in which direct interaction with ATC is necessary: in those cases, appropriate licensing would be mandatory for a standard radio communication by the UAV pilot/operator.

As far as a pilot's practical training is concerned, some countries require specific sessions and examinations; some other countries appear to leave the practical training to the UAV pilot or to the operator, once theoretical competence has been demonstrated.

Depending on specific national contexts and technical requirements established for a specific drone and/or its intended application, different obligations rest on the operator and/or the pilot.

In some cases, operators are expected to have and document an organizational, technical, and commercial infrastructure which must be adequate for the type of operations to be conducted.

National qualifications for small UAV pilots are generally non transferrable from one country to others without some sort of ad hoc clearance or validation examination.

2.4.3 UAV Registration/Validation

In many cases, professional UAVs are supposed to be known to the authorities[7]. This may consist in declaring the existence of the drone, its specific features and data, and its intended purpose; or it may consist in a complete validation process by the respective CAA, including more or less complex documentation. The general idea is that the authorities should be able to identify and trace the drone at any time, especially in the case of mishap or misconduct. Several solutions are in place and will possibly be further developed in the future. Typically, UAVs must bear plates with identification/registration codes; some regulations consider the application of in-flight identification devices (e.g., transponders). Flight recorders and other solutions are being tested to ensure that this univocal identification of

[7] e.g. Argentina (2015), Australia (2017), Bangladesh (2016), Belgium (2016), Botswana (2015), Brazil (2016), Chile (2015), China (2015), Cyprus (2015), France (2012), Germany (2017), Ghana (2016), Israel (2011), Italy (2016), Malaysia (2008), Mexico (2016), Nigeria (2015), Philippines (2015), Rwanda (2015), Slovakia (2015), Slovenia (2016), South Africa (2015), Spain (2016), Sri Lanka (2017), United Kingdom (2015, USA (2016), Vietnam (2008).

drones (as associated to their respective operators and pilots) is possible with increasing accuracy.

2.4.4 UAV Categorization

In most cases, national regulations seem to classify drones based on their weight[8]. This is pretty obvious from a safety point of view, but it may prove insufficient for some specific purposes (such as training requirements or operational scenario definitions). In fact, several additional criteria for UAV classifications are put in place by national CAAs. Some of these classifications are also based on "internal" features of the drone or its operational purposes[9]: equipment, deployment, etc. Some others relate to the operational scenarios, as it will be specified in the next section. In the end, drones appear to be occasionally subject to complex multi-aspect categorization which affect more or less directly the corresponding qualification requirements for the pilots and the operators.

2.4.5 Operational Scenarios

Drones can technically operate in virtually all available environments: limitations – sometimes severe – are defined in different countries and by individual CAAs within a certain country. The operational scenarios are the main discriminating element when it comes to deliver the drone to the field and launch it. It would be obvious to say that the more interference the operation causes to people and things (in terms of safety/security/privacy, etc.), the more restrictions should be in place. That is in fact what CAAs try to do; but the outcome is widely differentiated. Most regulations forbid to fly over densely populated areas and "crowds" or "people gathering" or "open-air assembly of people"[10]; what exactly the terms mean varies quite a bit from one CAA to the other. Georgia (2016) and Ireland (2015), for example, explicitly indicate 12 or more people to form a crowd and prohibit flight over them, United

[8] e.g. Argentina (2015), Australia (2017), Belgium (2016), Botswana (2015), Brazil (2016), Canada (2014), China (2015), Colombia (2015), Cyprus (2015), Czech Republic (2013), Denmark (2004), Dominican Republic (2015), Finland (2017), Germany (2017), Ghana (2016), Indonesia (2015), Ireland (2015), Italy (2016), Lithuania (2014), Mexico (2016), Nepal (2015), New Zealand (2015), Portugal (2016), Romania (2016), Rwanda (2015), Slovakia (2015), Slovenia (2016), Spain (2016), Sri Lanka (2017), Ukraine (2016), United Arab Emirates (n.d.), United Kingdom (2015), USA (2016).

[9] e.g. Austria (2015), China (2015), France (2017).

[10] e.g. Argentina (2015), Australia (2017), Bangladesh (2016), Belgium (2016), Brazil (2016), Chile (2015), Colombia (2015), Czech Republic (2013), Denmark (2004), Fiji (2016), France (2017), Germany (2017), Ghana (2016), Indonesia (2015), Italy (2016), Jamaica (2016), Malaysia (2008), Nepal (2015), Rwanda (2015), Slovakia (2015), Sri Lanka (2017), United Kingdom (2015), USA (2016).

Kingdom (2015) and Nepal (2015) indicate "more than 1,000 persons" for a 150 m mandatory separation. In most cases, a specific threshold number of people is not indicated.

Safety relevant areas include generally built zones[11]. An important practical variable is whether the zone is sparsely built or densely built. No clear quantitative threshold of density appears to be given in most regulations. Other areas of limited or forbidden access are those open to vehicle transit and various kinds of infrastructures. Although many regulations explicitly prohibit to fly over (more or less) inhabited areas, no-fly rules often include individual buildings, persons, vehicles (including vessels), and animals[12]. Besides specific and more or less severe restrictions, a general distinction can be inferred, from the documents we have examined, in the requirements for the use of UAVs in depopulated areas and for those in populated areas. A formal distinction appears to be made, for instance, in Italy (2016) regulation as it states that operations (not drones nor sites) should be divided into "critical" and "noncritical." This refers, respectively, to operations with specific hazard conditions and to others which can be considered unharmful even in the case of accidents.

Small drones are often limited in their permissible operational range. Widespread height limits above ground level are 120 and 150 m (i.e. roughly 400 ft and 500 ft) although allowed maximum heights can be as low as ~30 m (100 ft) (Nicaragua 2014) and permissible horizontal distances are variously assigned from 30 m (~100 ft) (Nicaragua 2014) to 1,000 m (~3,300 ft) (e.g. Lithuania 2014) and 1,500 m (~5,000 ft) (Colombia 2015), depending on various technical features and operational conditions.

Metric indications are obviously subject to further limitation due to other operational restrictions (e.g., presence of inhabited buildings). Another element, related to distance but more generally involved in operational limitations, is whether flights are to be performed in VLOS (Visual Line of Sight), EVLOS (Extended Visual Line of Sight), and BVLOS (Beyond Visual Line of Sight).

VLOS is the condition in which the remote pilot is able to see the drone. EVLOS is the condition (somewhat unspecified) by which the drone is in sight of a remote pilot either because the pilot is equipped with supporting personnel/systems that allow a visual contact with the drone or because the drone is flown in VLOS by a pilot who then hands it over to another pilot whose position allows a new VLOS condition[13]. BVLOS is the condition by which the drone is simply out of sight and is flown instrumentally[14]. Some CAAs mandate that BVLOS could be performed only by drones fitted with "sense-and-avoid" devices (not yet standardized) so as to ensure separation from other air traffic. The regulatory management of potential risks based on operational scenario and technical features of drones is one of the most differentiated aspects of UAV-related legislation.

---

[11] e.g. Bangladesh (2016), Canada (2014), Denmark (2004), Ghana (2016), Italy (2016, 2017), Lithuania (2014), Slovakia (2015), United Kingdom (2015).

[12] e.g. Azerbaijan (2015), Bangladesh (2016), Canada (2014), Cyprus (2015), France (2017), Ghana (2016), Italy (2016), Lithuania (2014), Malaysia (2008), Mexico (2016), Slovakia (2015), Sri Lanka (2017).

[13] e.g. Brazil (2016), Italy (2016, 2017).

[14] e.g. Austria (2015), Brazil (2016), China (2015), Ghana (2016), Italy (2016), USA (2016).

An explicit distinction is made, by some CAAs between nonparticipating persons and persons belonging to the organization operating the UAV (e.g., Bangladesh, Italy).

2.4.6 Flight Areas and Limitations

Most regulations indicate no-fly areas, although some of them are explicitly stated and some others are not. Usual forbidden areas – unless specific clearances are granted – are airports and surroundings. A safe distance is often required, essentially to prevent any interference between manned air traffic and the drone[15].

For similar reasons, flying drones within emergency areas may be restricted to authorized operations, as uncoordinated UAV flight could interfere with rescue air traffic[16].

In general, areas of cities or territories can be specifically restricted to drones[17], along with other areas which are already forbidden to usual air traffic (e.g., prohibited, restricted, and dangerous zones); given that drones can be deployed close to institutionally, politically, or strategically relevant buildings, additional no-fly zones for drones are variously defined, either permanently or provisionally.

2.4.7 Privacy

Although privacy is becoming a major concern, only a few national CAAs are explicitly including it in their regulations. This might be due to the fact that the right to privacy and sanctions against violations are covered in other parts of the respective national legislations. A most basic requirement is that anytime a survey or observation flight is performed, permission must be granted by the owner or the manager of the observed area, or by other people involved[18]. Some regulations specify that recording and preservation of data (images, video footage, etc.) acquired by the drone must be authorized specifically by the owner of the privacy right. As an alternative, the data itself must be acquired in such a way so as to prevent personal or privacy-relevant data to be included in the process. This is the

[15] e.g. Australia (2017), Belgium (2016), Brazil (2016), France (2017), Germany (2017), Italy (2016, 2017), Jamaica (2016), Malaysia (2008), Nepal (2015), Rwanda (2015), Sri Lanka (2017).

[16] e.g. Australia (2017), Chile (2015), Germany (2017).

[17] e.g. Argentina (2015), Bangladesh (2016), China (2015), Denmark (2004), France (2017), Georgia (2016), Germany (2017), Ghana (2016), India (2016), Indonesia (2015), Italy (2016), Mexico (2016), Nepal (2015), Portugal (2016), Rwanda (2015), Slovakia (2015).

[18] e.g. Cyprus (2015), Germany (2017), Ghana (2016), Jamaica (2016).

case, for instance, when images are taken at a resolution that does not allow recognition of accidentally viewed persons[19].

2.5 Small Drones and Geographic Survey: Not Everything, Just What They Are Really Good At

Gianluca Casagrande

A thought-provoking work by Jackman (2016) considers the rethorics of "possibility" and "inevitability" commonly associated to the use of UAVs. In observing commercial narratives and the quick evolution of technologies, small drones are often associated with the idea that virtually any kind of need can be fulfilled by emerging products. From some points of view, such an idea may prove inaccurate.

On the one hand, technology is still at a relatively early stage of its development: how it will end up being integrated with practical uses and regulatory frameworks is still to be fully understood. On the other hand, real-life applications bind the user to deal with available systems in actual scenarios. Many variables involve their effective capability of providing the intended information. In the previous sections, we have discussed aspects of geographic relevance such as scales and area coverage. We have also discussed how different drone implementations play a role in those aspects. We have then discussed field deployment and the general concept of accessibility; finally, we proposed an overview of typical legal issues.

Correct mission planning involves a full understanding of the UAV overall performance. Most of all, however, it involves the ability to recognize whether the drone is effective for the intended purpose *or not*.

A thorough examination of pros and cons about the use of drones for reification surveys is presented by Campbell and Katz (2016). The authors appropriately observe that small drones are suitable for some kinds of applications and just unsuitable for others.

Limited payload capabilities restrict the set of usable instrumentation to sufficiently miniaturized systems. This limits the applicability of other, more effective technologies until they become sufficiently small and sufficiently cheap to be installed on drones.

Current drones are able to cover and document larger areas than those normally walked by an individual surveyor in corresponding time; however, for surfaces above 10,000 acres (~4,000 ha or ~40 km^2) according to the cited authors, manned aircraft become more cost-efficient. In cartographic and photogrammetric applications, the best accuracy a commercial drone can normally provide is in the range of a few centimeters, adequate to topographic mapping and sometimes for architecture as well, but yet below the performance of other traditional techniques. The popular photographic and photogrammetric procedures, around which drones are booming, have a

[19] e.g. Italy (2017), art. 34.

typical limitation when the terrain surface is beneath dense land cover. Airborne laser scanning can be more efficient in discriminating surfaces, but its adoption on board small drones is still relatively rare and severely affects affordability.

Besides these general considerations, it is worth noting that the actual capability of small UAVs to perform the job intended by users depends heavily on the nature of the available technology. An example – a meaningful one, in our opinion – is drawn from a case in our own work on the field.

The purpose of the activity was a specific comparative research about small drones in close-range archaeological photogrammetry. Tests were conducted in early 2016 at Otricoli (Terni, Italy).

The idea was to conduct a comparison between two different minimal multi-rotor UAVs, fitted with built-in cameras. They were later to be compared to a standard small UAV equipped with an action camera. The goal was to assess the respective aerophotogrammetric capabilities. The three drones were used to perform an acquisition survey in a real archaeological site. The criteria for surveys were chosen according to probable typical profiles and contexts for minimal-drone operations:

1. Expeditive data acquisition and processing.
2. Close-range photogrammetry of relatively small areas and buildings.
3. Limited field support.
4. Drones were supposed to be flown manually or in GPS-assisted manual mode and in free flight (i.e., not programmed) trajectories.

Images were acquired, for all three drones, by sectioning video clips into individual frames. This was considered to present a credible worst-case scenario. Indeed, in the particular case, individual frames bore no exif nor positional data. A standard photogrammetric software (Agisoft PhotoScan) was used for processing. Another standard professional software (PhotoModeler Scanner) was used to compare results and to perform additional computations in order to assess accuracy.

Drones were flown in very close-range passes, at no more than 5 m distance from the target objects, i.e., vertical archaeological remains. A system of ground control points had been established and measured by TST in order to evaluate accuracy of systems under comparison. Weather was overcast, and wind was calm.

The three drone systems (UAV + cameras) operated as expected, and errors were within acceptable limits. Drone "A" (weight 0.13 kg or ~0.29 lb) was the most financially and operationally accessible and the easiest to deploy on the field. It was rather challenging as far as piloting was concerned and provided no assistance in flight control; hence all operations were necessarily in fully manual mode. This UAV was also the most sensitive to turbulence due to minimal weight and relatively high drag. It did not provide any stabilization to the camera. The obtained 2 Mpx images were therefore affected by frequent blurring and distorsions. A very time-consuming process was the manual selection of images which were then to be fed into the photogrammetric software. No adequate automatic selection system was available. For these reasons, careful examination was to be performed in order to preselect a series of sufficiently sharp and geometrically correct images.

In summary, the most accessible flying platform was indeed the most difficult one in terms of sensor quality and required the longest time for data processing. The intermediate drone, UAV "B" (weight 0.29 kg or ~0.64 lb), whose accessibility was still high, was a semiprofessional platform with a 3-axis digitally gyrostabilized camera; images (full-HD video frames) had very good geometric quality but only a fairly good chromatic quality, and the sensor suffered minimal but perceivable blurring when oriented to the extreme of the rotation area. Yet, image selection was easy and processing proved straightforward. The largest drone, UAV "C" (still a small UAV but a 2.3 kg or ~5 lb hexacopter), employed a standard action camera whose use is typical in archaeological surveys. The camera was mounted on a 3-axis mechanically gyrostabilized gimbal. Its full-HD images were consistently sharp, and colors were at state-of-the-art quality. Yet, the UAV + camera system was the less accessible and the less field deployable; it was also, by far, the most financially expensive and the most complex to manage from a regulatory point of view.

In the discussed case, the true sense of the comparison stood in the accuracy assessment. The average error was found to be 2.4 cm for UAV "A," 1.2 cm for UAV "B," and even 1.4 cm for UAV "C." If we discount the decimals, we can state a point that – we believe – is the essential one. In the course of a close-range photogrammetric acquisition, drones "B" and "C" had substantially the same performance as far as the purpose of the operation was concerned: both were suitable for 1:20 or 1:50 scale representations. Drone "B" costed only about 50% of drone "C." Drone "C" was more cumbersome to deploy, and it could be operated only with many more regulatory restrictions. Drone "A" costed just about 4% of drone "C" and provided a significantly larger accuracy error, but the overall error was still perfectly acceptable for expeditive archaeological surveys (i.e., suitable for 1:100 scale representations).

A few days after the above-described activity, drones "B" and "C" were also used in another kind of comparative situation. Although it was not planned and intended to be a strict comparison, it turned out to be so de facto. The two machines were now used to take simple, panoramic aerial views of a natural and archaeological area in Tor Caldara (Rome, Italy). In both cases, footage was taken to provide some qualitative description of the site, its cultural heritage, and natural assets. At the time of flights, weather was clear, and a constant moderate wind was blowing across the observation area.

The difference in the quality of images and the kinematic capabilities of the two drones gave a clear indication of their value as systems for that particular operation. Drone "C" provided state-of-the-art footage, with a perfectly stabilized video (thanks to the quality of sensor and excellent stabilization system). It also allowed for smooth and fast spatial transitions from one view to the next one. Due to the inherent mass and the quite large amount of available motor power, the drone could perform all the intended maneuvers without revealing any evident disturbance by wind. Wide trajectories at 200–500 m (~600 to ~1,600 ft) distance from the control station could be performed safely, thanks to the 10–15 min endurance time. The stable and reliable real-time video telemetry communication channel allowed to perform FPV-assisted maneuvers.

Drone "B," a lightweight quadcopter with about 8 min endurance, was able to produce perfectly stabilized images whose overall chromatic quality appeared to be less appreciable than the other system. Due to its smaller dimensions and modest absolute power, the drone could perform stable panoramic view transitions by maneuvering around the area; however, because of the lower airspeed and the presence of the wind, motion respective to the ground appeared to be variable. Maximum distance was intentionally contained within 150 m (~500 ft), also to reduce occasional degradation of the FPV signal. Moreover, in one occasion during the flight, at a distance of about 150 m from the ground station, drone "B" suffered a temporary disconnection of the flight control and data link channels, presumably due to a temporary obstruction of the direct line of sight between the transmitter and the receiver as the drone was passing "behind" a building. The connection was promptly reestablished by changing the position of the control station. The drone, which had stopped its navigation and was holding in hovering, resumed controlled flight. This type of inconvenience never occurred to drone "C."

For the sake of completeness, it is worth mentioning that drone "A," although briefly, was indeed flown on that same day. It produced low-quality panoramic images which were impossible to use for professional video documentation, albeit adequate for general observation of the site in its basic features. A drawback in that type of application would be that drone "A," unlike the other two UAVs, could not downlink any FPV signal. For this reason, the recorded view could not be evaluated by the pilot during shooting. Furthermore, in other situations, drone "A" had proved quite prone to signal loss whenever, at distance of 50 m (~160 ft), obstacles should interfere with transmitter-receiver alignment.

It is additionally worth noting that drones "B" and "C" were both programmable, depending on operational conditions, to remain hovering or to navigate back to homebase and auto-land in case of signal loss. Drone "A" had only a single fail-safe functionality: it switched to auto-landing mode and descended to the ground, on the spot, immediately after signal loss. This would have practical consequences in the likely case the drone would have landed in a place difficult or impossible to reach for the pilot.

In summary, during the second (informal) test, we had taken the three drones into another operational context, i.e., in wider area video documentation. Drone "A" showed poor performance and unsatisfactory reliability (though retaining its perfectly inoffensive nature). This suggested that its use was operationally inappropriate in that type of application, unless for extreme cases in which no other solutions were viable. Drone "B" proved adequate for semiprofessional quality footage, and drone "C" proved fully adequate for professional documentation.

References

Adams Z, Benedict M et al (2013) Design, development, and flight test of a small-scale cyclogyro UAV utilizing a novel cam-based passive blade pitching mechanism. Int J Micro Air Veh 5(2):145–162

Aden ST, Bialas JP et al (2014) Low cost infrared and near infrared sensors for UAVs. Int Arch Photogramm Remote Sens Spat Inf Sci. XL-1. doi:10.5194/isprsarchives-XL-1-1-2014

Argentina (2015) Administracion Nacional de Aviación Civil – Proyecto de reglamento provisional de los vehiculos aéreos no tripulados. Available via http://www.anac.gov.ar/anac/web/uploads/normativa/resoluciones/anexo-resolucion-041-2015.pdf. Accessed 20 Apr 2017

Arteaga Botello N (2016) Política de la verticalidad: drones, territorio y población en América Latina. Región Soc 65:263–292

Australia (2017) Australian government – Civil Aviation Safety Authority. Advisory circular. AC 101–01 remotely piloted aircraft systems – licensing and operations. Available via https://www.casa.gov.au/files/101c01pdf. Accessed 31 Aug 2017

Austria (2015) Aviation Agency – Operation of Remotely Piloted Aircraft Systems (RPAS). Available via https://www.austrocontrol.at/en/aviation_agency/licenses__permissions/flight_permissions/rpas. Accessed 20 Apr 2017

Azerbaijan (2015) Republic of Azerbaijan – State Civil Aviation Administration. Unmanned aircraft operations. Available via http://www.caa.gov.az/index.php?option=com_k2&view=item&id=297:unmanned-aircraft-operations&Itemid=174&lang=en. Accessed 31 Aug 2017

Bangladesh (2016) Civil Aviation Authority, Bangladesh. Circular. Regulations for Operating Remotely Piloted Aircraft Systems (RPAS). Available via http://www.caab.gov.bd/corporate/rpas.pdf. Accessed 20 Apr 2017

Belgium (2016) 10 Avril 2016. Arrêté royal relatif à l'utilisation des aéronefs télépilotés dans l'espace aérien belge. Available via http://www.ejustice.just.fgov.be/cgi_loi/change_lg.pl?language=fr&la=F&cn=2016041002&table_name=loi. Accessed 20 Apr 2017

Beloev IH (2016) A review on current and emerging application possibilities for unmanned aerial vehicles. Acta Technol Agric 19(3):70–76. doi:10.1515/ata-2016-0015

Bendea H, Boccardo P et al (2008) Low cost UAV for post-disaster assessment. Int Arch Photogramm Remote Sens Spat Inf Sci XXXVII B8:1373–1379

Birtchnell T, Gibson C (2015) Less talk more drone: social research with UAVs. J Geogr High Educ 39(1):182–189. doi:10.1080/03098265.2014.1003799

Botswana (2015) Civil Aviation Authority of Botswana – legal requirements for operating remotely piloted aircraft in Botswana

Bracken-Roche C (2016) Domestic drones: the politics of verticality and the surveillance industrial complex. Geogr Helv 71:167–172. doi:10.5194/gh-71-167-2016

Brazil (2016) Ministério da Defesa – Comando da Aeronáutica. Tráfego Aéreo. ICA 100–40 Sistemas de aeronaves remotamente pilotadas e o acesso ao espaço aéreo brasileiro. Available via https://www.decea.gov.br/static/uploads/2015/12/Instrucao-do-Comando-da-Aeronautica-ICA-100-40.pdf. Accessed 20 Apr 2017

Cameroon (2016) Republique du Cameroun. Autorité Aeronautique – Instruction n° 12/C/CCAA/DG/DCRCA du 18 novembre 2016 fixant les modalités de demande d'autorisation d'utilisation d'aéronefs télépilotés dans l'espace aérien et sur le territoire camerounais pour des opérations ponctuelles de travail aérien. Available via http://www.dasis.ccaa.aero/documents/numérisation0002.pdf. Accessed 31 Aug 2017

Campbell L, Katz D (2016) The drone as a tool. What drones do well and what they don't. Point of beginning. October: 26–29

Canada (2014) Transport Canada. Review and processing of an application for a special flight operations certificate for the operation of an unmanned air vehicle (UAV) system. Available via https://www.tc.gc.ca/media/documents/ca-standards/(SI)_No._623-001_2_en.pdf. Accessed 20 Apr 2017

Carrivick JL, Smith MW et al (2013) Developments in budget remote sensing for the geosciences. Geol Today 29(4):138–143. doi:10.1111/gto.12015

Casagrande G, Salvatori F et al (2015) A "Low-tech" approach as a strategy for effective UAV surveying of landscape and cultural heritage. Experiences from the lab and the field. Documenti Geografici 2:33–65. doi:10.19246/dg.v0i2.80

Chad (2014) Autorité de l'Aviation Civile – RAT 02 Règles de l'air. Available via http://adac-tchad.org/wp-content/uploads/2015/05/RAT-02-REGLES-DE-LAIR.pdf. Accessed 20 Apr 2017

Chile (2015) Chile – Direccion General de Aeronautica Civil. DAN 151 Operaciones de aeronaves pilotadas a distancia (RPAS) en asuntos de interés público, que se efectúen sobre áreas pobladas. Available via https://www.dgac.gob.cl/portalweb/rest-portalweb/jcr/repository/collaboration/sites%20content/live/dgac/categories/normativas/normas DAN/documents/DAN_151–20150413.pdf. Accessed 20 Apr 2017

China (2015) 编 号:AC-91-FS-2015-31 中 国 民 用 航 空 局 飞 行 标 准 司 -轻小无人机运行规定 (试行). Available via http://www.caac.gov.cn/XXGK/XXGK/GFXWJ/201601/P020160126526845399237.pdf. Accessed 20 Apr 2017

Colombia (2015) Aeronáutica Civil de Colombia – Requisitos Generales de Aeronavegabilidad y Operaciones Para RPAS (Numeral 4.25.8.2). Available via http://www.aerocivil.gov.co/autoridad-de-la-aviacion-civil/certificacion-y-licenciamiento/Documents/PROYECTO%20BORRADOR%20CIRCULAR%20RPAS.pdf. Accessed 20 Apr 2017

Côte d'Ivoire (2017) Autorité Nationale de l'Aviation Civile de Côte d'Ivoire – Decision n° 00003436 / ANAC /DSV/DSNAA/DTA Portant Règlement Aéronautique de Côte d'Ivoire relatif aux aéronefs télépilotés RACI 3009 Available via: http://www.anac.ci/anac/web/docs/RACI%203009.pdf. Accessed 31 Aug 2017

Croatia (2015) Pravilnik o Sustavima Bespilotnih Zrakoplova Available via http://narodne-novine.nn.hr/clanci/sluzbeni/2015_05_49_974.html. Accessed 31 Aug 2017

Croatia (2006) Ministry of the Sea, Tourism, Transport and Development – ordinance on rules of the air. Available via http://www.mvep.hr/zakoni/pdf/353.pdf. Accessed 20 Apr 2017

Cyprus (2015) Cyprus government gazette – Civil Aviation Act 2002 (as amended). Ministerial decree issued in accordance with Articles 5 (1) and 260. Available via http://www.mcw.gov.cy/mcw/dca/dca.nsf/All/BF026F4CECA35180C2257DB30030A750/$file/UAVS%20Decree%20402-2015-english%20tanslation%20last%20revision%20jan%202016%20(3).pdf

Czech Republic (2013) Civil Aviation Authority Czech Republic. Unmanned aircraft regulatory framework in the Czech Republic. Available via http://jarus-rpas.org/sites/jarus-rpas.org/files/czech_uas_regulatory_framework_en.pdf. Accessed 20 Apr 2017

Denmark (2004) Civil Aviation Administration – Denmark – BL 9–4 regulations on unmanned aircraft not weighing more than 25 kg. Available via http://selvbetjening.trafikstyrelsen.dk/civilluftfart/Dokumenter/English/Rules%20and%20Regulations/Regulations%20for%20Civil%20Aviation%20(BL)/BL%2009-series/BL%209-4,%203%20edition.pdf. Accessed 20 Apr 2017

Dominican Republic (2015) Instituto Dominicano de Aviación Civil – Resolución No. 008-2015 que regula el uso y operacion de los sistemas de aeronaves pilotadas a distancia (RPAS) o drones en el territorio nacional

Fiji (2016) Civil Aviation Authority of Fiji – Fiji aeronautical information circular. Operation of Remotely Piloted Aircraft Systems (aka RPAS, U.A.V. or Drones). Available via http://www.caaf.org.fj/index.php?option=com_jdownloads&Itemid=61&task=finish&cid=891&catid=82. Accessed 20 Apr 2017

Finland (2017) Finnish Transport Safety Agency, Use of Remotely Piloted Aircraft and Model Aircraft, 23 December 2016, Available via https://www.trafi.fi/filebank/a/1483970125/4a6ac53bf4b1cb434d7f85a15f36dde0/23661-OPS_M1-32_RPAS_2016_eng.pdf. Accessed 31 Aug 2017

Finn RL, Wright D (2012) Unmanned aircraft systems: surveillance, ethics and privacy in civil applications. Comput Law Secur Rev 28(2):184–194. doi:10.1016/j.clsr.2012.01.005

France (2017) Arrêté du 17 décembre 2015 relatif à l'utilisation de l'espace aérien par les aéronefs qui circulent sans personne à bord, Version consolidée au 13 avril 2017, Available via https://www.legifrance.gouv.fr/affichTexte.do?cidTexte=JORFTEXT000031679868&dateTexte=20170413. Accessed 31 Aug 2017

Fryskowska A, Kedzierski M et al (2016) Calibration of low-cost RGB and NIR UAV cameras. Int Arch Photogramm Remote Sens Spat Inf Sci XLI-B1:817–821. doi:10.5194/isprsarchives-XLI-B1-817-2016

Gabon (2014) Republique Gabonaise – Reglement Aeronautique Gabonais. Regles de l'air RAG 7.1. Available via http://anacgabon.org/fr/images/pdf/rag7.1.pdf. Accessed 20 Apr 2017

Georgia (2016) Georgian civil aviation authority – drones. Available via http://www.gcaa.ge/eng/RPAS.php. Accessed 20 Apr 2017

Germany (2017) Bundesgesetzblatt Jahrgang 2017 Teil I Nr. 17, ausgegeben zu Bonn am 6. April 2017, Verordnung zur Regelung des Betriebs von unbemannten Fluggeräten, Bundesanzeiger Verlag, pp. 683–688. Available via https://www.bgbl.de/xaver/bgbl/start.xav?startbk=Bundesanzeiger_BGBl&start=%2F%2F*[%40attr_id=%27bgbl117s0683.pdf%27]#__bgbl__%2F%2F*%5B%40attr_id%3D%27bgbl117s0683.pdf%27%5D__1504446223278. Accessed 31 Aug 2017

Ghana (2016) Ghana Civil Aviation (Safety) Directives, 2016 – part 28. Remotely Piloted Aircraft Systems. Available via http://www.gcaa.com.gh/extweb/images/final%20rpas%20safety%20directive.pdf. Accessed 20 Apr 2017

Giges NS (2016) Printing UAVs on demand. Mechanical Engineering, November 2016: 12 and 15

Hilton RK, Shaw JM (2016) I see you. Drones, privacy torts and insurance claims. Claims Magazine, December: 21–23

Hristov GV, Zahariev PZ, Beloev IH (2016) A review of the characteristics of modern unmanned aerial vehicles. Acta Technol Agric 19(2):33–38. doi:10.1515/ata-2016-0008

India (2016) Government of India – Office of the Director General of Civil Aviation. Technical Centre, Opp. Safdarjung Airport, New Delhi – Air Transport Circular XX of 2016. Available via http://dgca.nic.in/misc/draft%20circular/AT_Circular%20-%20Civil_UAS(Draft%20April%202016).pdf. Accessed 20 Apr 2017

Indonesia (2015) Peraturan Menteri Perhubungan Republik Indonesia – Nomor Pm 180 tahun 2015 tentang pengendalian pengoperasian sistem pesawat udara tanpa awak di ruang udara yang dilayani Indonesia. Available via http://jdih.dephub.go.id/assets/uudocs/permen/2015/PM_180_Tahun_2015.pdf. Accessed 20 Apr 2017

Ireland (2015) Irish Aviation Authority small unmanned aircraft (drones) and rockets order, 2015. Available via http://www.irishstatutebook.ie/eli/2015/si/563/made/en/pdf. Accessed 20 Apr 2017

Israel (2011) DIRECTIVE הוראה – רישום, רישוי, תיעוד ושינויים לכטב"ם. Available via http://caa.gov.il/index.php?option=com_docman&view=download&alias=5209–2011-28&category_slug=2015–10–13-06-38-50-7&Itemid=669&lang=he. Accessed 20 Apr 2017

Italy (2016) ENAC Regolamento (Regulation) – Mezzi aerei a pilotaggio remoto (Remotely Piloted Aerial Vehicles). Available via http://www.enac.gov.it/repository/contentmanagement/information/n1220929004/regulation_rpas_issue_2_rev%202_eng.pdf. Accessed 20 Apr 2017

Italy (2017) ENAC Regolamento – Mezzi aerei a pilotaggio remoto. Edizione 2 del 16 luglio 2015 – Emendamento 3 del 24 marzo 2017. Available via http://www.enac.gov.it/repository/ContentManagement/information/N122671512/Regolamento_APR_Ed_2_Emend_3.pdf. Accessed 5 May 2017

Jackman AH (2016) Rhetorics of possibility and inevitability in commercial drone tradescapes. Geogr Helv 71:1–6. doi:10.5194/gh-71-1-2016

Jamaica (2016) Jamaica Civil Aviation Authority – flight safety notification general. Operation of unmanned aerial vehicles. Available via http://jcaa.gov.jm/wp-content/uploads/2016/04/FSN-GN-2015-01.pdf. Accessed 20 Apr 2017

Lelong CCD, Burger P et al (2008) Assessment of unmanned aerial vehicles imagery for quantitative monitoring of wheat crop in small plots. Sensors 8(5):3557–3585. doi:10.3390/s8053557

Lithuania (2014) Civilinės Aviacijos Administracijos Direktorius – Įsakymas dėl Bepiločių Orlaivių Naudojimo Taisyklių Patvirtinimo. Available via https://e-seimas.lrs.lt/portal/legalAct/lt/TAD/7bb14b10849f11e39d2dc0b0e08d5f21. Accessed 20 Apr 2017

Malaysia (2008) Malaysia – Aeronautical Information Services Department of Civil Aviation. Unmanned aerial vehicle (UAV) operations in Malaysian airspace. Available via http://aip.dca.gov.my/aip%20pdf%20new/AIC/AIC%20200804.pdf. Accessed 20 Apr 2017

Mauritania (2014) Agence Nationale de l'Aviation Civile – Reglement Technique Aeronautique RTA – 02. Regles de l'air. Available via http://www.anac.mr/IMG/pdf/RTA022014.pdf. Accessed 20 Apr 2017

Mexico (2016) Dirección General de Aeronáutica Civil – Circular Obligatoria que establece los requerimientos para operar un sistema de aeronave pilotada a distancia (RPAS). Available via http://www.sct.gob.mx/fileadmin/DireccionesGrales/DGAC-archivo/modulo3/co-av-23-10-r3.pdf. Accessed 20 Apr 2017

Nepal (2015) Civil Aviation Authority of Nepal – flight operations directives No. 07. Available via https://irevolution.files.wordpress.com/2015/05/caa_nepal-uav-regulations.pdf. Accessed 20 Apr 2017

Netherlands (2015) Inspectie Leefomgeving en Transport – Ministerie van Infrastructuur en Milieu. Flights with remotely piloted aircraft systems (RPAS) – so called drones – in the Netherlands. Available via https://english.ilent.nl/themes/rpas---remotely-piloted-aircraft-systems-drones. Accessed 20 Apr 2017

New Zealand (2015) Civil Aviation Authority of New Zealand – part 101 CAA consolidation. Gyrogliders and parasails, unmanned aircraft (including balloons), kites, and rockets – operating rules. Available via https://www.caa.govt.nz/rules/Rule_Consolidations/Part_101_Consolidation.pdf. Accessed 20 Apr 2017

Nicaragua (2014) Instituto Nicaragüense de Aeronáutica Civil – Nota de Prensa. Available via http://www.inac.gob.ni/wp-content/uploads/2014/11/Nota-de-Prensa-INAC.pdf. Accessed 20 Apr 2017

Nigeria (2015) Federal Republic of Nigeria – Official Gazette. Available via http://www.ncaa.gov.ng/media/1273/aviation-part-10.pdf. Accessed 20 Apr 2017

Ogden LE (2013) Drone ecology. Bioscience 63(9):776. doi:10.1525/bio.2013.63.9.18

Pérez M, Agüera F, Carvajal F (2015) Low cost surveying using an unmanned aerial vehicle. Int Arch Photogramm Remote Sens Spat Inf Sci XL-1(W2):311–315. doi:10.5194/isprsarchives-XL-1-W2-311-2013

Philippines (2015) Civilian Aviation Authority of the Philippines. Available via http://caap.gov.ph/index.php/downloads/send/78–2015/293-mc-29-15-amendments-to-pcar-part-2-4-11. Accessed 20 Apr 2017

Portugal (2016) Autoridade Nacional da Aviação Civil Regulamento n. 1093/2016 – Condições de operação aplicáveis à utilização do espaço aéreo pelos sistemas de aeronaves civis pilotadas remotamente ("*Drones*"). Available via https://dre.pt/application/file/105366569. Accessed 20 Apr 2017

PwC (2016) Clarity from above, PwC global report on the commercial applications of drone technology, May 2016. Available via https://www.pwc.pl/pl/pdf/clarity-from-above-pwc.pdf. Accessed 19 Apr 2017

Romania (2016) Romanian CAA Directivă de navigabilitate – Emiterea certificatelor de identificare pentru aeronavele fără pilot la bord (RPAS). Available via http://www.caa.ro/pdf/Directiva%20identificare%20UAV.pdf. Accessed 20 Apr 2017

Rwanda (2015) Civil aviation (unmanned aircraft system (UAS) regulations 2015 – arrangement of regulations). Available via http://www.caa.gov.rw/fileadmin/templates/Regulatory_Services/Annex_to_the_Ministerial_regulations_on_UAS_Regulations_17_March_2016.pdf. Accessed 20 Apr 2017

Slovakia (2015) Dopravný Úrad. Transport Authority – Rozhodnutie č. 1/2015 z 19.08.2015, ktorým sa určujú podmienky vykonania letu lietadlom spôsobilým lietať bez pilota vo vzdušnom priestore Slovenskej republiky. Available via http://nsat.sk/wp-content/uploads/2014/08/DU_RPAS-merged.pdf. Accessed 20 Apr 2017

Slovenia (2016) Uredbo o sistemih brezpilotnih zrakoplovov. Available via https://e-uprava.gov.si/drzava-in-druzba/e-demokracija/predlogi-predpisov/predlog-predpisa.html?id=7217. Accessed 20 Apr 2017

Smith P (2013) Invasion of the drones? N Y Times Upfront 22:10–11

South Africa (2015) South African Civil Aviation Authority – Remotely piloted aircraft systems (Part 101) regulations workshop. RPAS regulations (Part 101). Available via http://www.caa.co.za/Documents/RPAS/Part%20101%20-%20RPAS%20Workshops.pdf. Accessed 20 Apr 2017

Spain (2016) Ministerio de Fomento – Ministerio de Defensa – Real Decreto por el que se regula la utilización civil de las aeronaves pilotadas por control remoto, y se modifica el Real Decreto 552/2014, de 27 de junio, por el que se desarrolla el Reglamento del aire y disposiciones operativas comunes para los servicios y procedimientos de navegación aérea y se modifica el Real Decreto 57/2002, de 18 de enero, por el que se aprueba el Reglamento de Circulación Aérea. Available via https://www.fomento.es/NR/rdonlyres/63ECAE3A-B29E-45A7-A885-D314153883EE/139826/RDRPAS27102016.pdf. Accessed 20 Apr 2017

Sri Lanka (2017) Civil Aviation Authority of Sri Lanka – Requirements for Operation of Pilotless Aircraft (Unmanned Aerial Vehicles Unmanned Aircraft Systems/Remotely Piloted Aircraft/Drones). Available via http://www.caa.lk/images/stories/pdf/Drones/SLCA_IS053_1st_Edition_Rev02_10JAN2017.pdf. Accessed 20 Apr 2017

Togo (2015) République du Togo Ministère chargé de L'Aviation Civile – Réglements Aéronautiques Nationaux du Togo RANT 02 Régles de l'air. Available via http://anac-togo.tg/RANTS/RANT_02/RANT_02_regles_de_lair.pdf. Accessed 20 Apr 2017

Ukraine (2016) Available via https://www.loc.gov/law/help/regulation-of-drones/ukraine.php#_ftn25. Accessed 20 Apr 2017

United Arab Emirates (n.d.) General Civil Aviation Authority. Available via https://www.gcaa.gov.ae/en/Pages/UASRegistration.aspx. Accessed 20 Apr 2017

United Kingdom (2015) Civil Aviation Authority – Unmanned Aircraft System Operations in UK Airspace – guidance. Available via http://publicapps.caa.co.uk/docs/33/CAP%20722%20Sixth%20Edition%20March%202015.pdf. Accessed 20 Apr 2017

United States of America (2016) Federal Aviation Administration (FAA) and Office of the Secretary of Transportation (OST), Department of Transportation (DOT) - Operation and Certification of Small Unmanned Aircraft Systems. Available via: https://www.federalregister.gov/documents/2016/06/28/2016-15079/operation-and-certification-of-small-unmanned-aircraft-systems. Accessed 31 Aug 2017

Vietnam (2008) The Government Socialist Republic of Viet Nam – decree on management of unmanned aircraft and ultralight aircraft. Available via http://moj.gov.vn/vbpq/en/Lists/Vn%20bn%20php%20lut/View_Detail.aspx?ItemID=2867. Accessed 20 Apr 2017

Villa TF, Gonzalez F et al (2016) An overview of small unmanned aerial vehicles for air quality measurements: present applications and future prospectives. Sensors 16(7):1072. doi:10.3390/s16071072

Wang Y, Xia H et al (2016) Flying eyes and hidden controllers: a qualitative study of people's privacy perceptions of civilian drones in the US. Proc Priv Enhancing Technol 3:172–190. doi:10.1515/popets-2016-0022

Watts AC, Ambrosia VG, Hinkley EA (2012) Unmanned Aicraft Systems in remote sensing and scientific research: classification and considerations of use. Remote Sens 4(6):1671–1692. doi:10.3390/rs4061671

Yun M, Kim J et al (2013) Feasibility of employing a smartphone as the payload of a photogrammetric UAV system. ISPRS J Photogramm Remote Sens 79:1–18. doi:10.1016/j.isprsjprs.2013.02.001

Zhao B, Hellwich O et al (2015) Employing smartphone as on-board navigator in unmanned aerial vehicles: implementation and experiments. Ind Robot: Int J 42(4):306–313. doi:10.1108/IR-01-2015-0016

Chapter 3
Opportunities

Gianluca Casagrande

Abstract This chapter presents some aspects of how drones are expanding the potential of aerial geographic observation. General basics are summarized in the first section (3.1). Some development lines of the use of small UAV in geography-related remote sensing are then discussed from a theoretical point of view (3.2). The following section (3.3) presents a series of current applications in reference to selected works from the scientific and divulgation literature and news from the Web. The adopted point of view on the topics is centered on the data acquisition capability of small UAVs. Such an approach highlights the fact that currently available technologies can possibly be used to fulfill several and sometimes thematically distant purposes in geographic observation. The last section (3.4) is devoted to a general discourse on the potential of small drones toward the exploration of historic and cultural landscapes, from archaeology to a more humanistic-oriented interpretation of present-day places.

Keywords UAV • Aerial images • Aerial survey • Geographic observation • UAV-based environmental observation • Landscape monitoring • UAV-based risk management • aerial archaeology • Cityscape

3.1 Geographic Observation

Human beings live in several types of geographical spaces. While really "anaecumenical" regions do not exist any longer, there are remote areas where access may prove impervious and yet important for various reasons. UAVs are becoming effective tools in those contexts especially for research.

Naturally, drones are mostly used in spaces which are directly involved by human presence and action.

In this part of the book, we discuss aspects concerning two important domains: environment, on the one hand, and anthropized space on the other hand.

G. Casagrande (✉)
Geographic Research and Application Laboratory, European University of Rome, Rome, Italy
e-mail: gianluca.casagrande@unier.it

G. Casagrande et al. (eds.), *Small Flying Drones: Applications for Geographic Observation*, DOI 10.1007/978-3-319-66577-1_3

The physics of nature- and human-generated phenomena produces effects which can be spotted and evaluated by applying similar tools in similar ways: drones are becoming increasingly able to “see” those effects, and let the user know.

The parallel development of miniaturized, very powerful cameras and robust data processing systems allowed for highly efficient and very affordable geometrical reconstruction of physical objects. This opened a new phase in 3D modeling of individual geographic features or entire areas.

Many other types of sensors appeared in the past decades; this allowed to activate a massive application of new detection and measurement techniques. The fact that radiations in electromagnetic bands other than visible light could be explored was an important achievement in satellite and manned airborne Earth observation. When miniaturization made it possible to install those sensors on small UAVs at a reasonable cost, their use got an important boost in terms of flexibility and level of detail.

According to their functional principles, sensors can be purely passive or active. Passive sensors operate by detecting and measuring radiation they receive from an external source. This is the case, for instance, for photocameras, thermal cameras, tetracams, etc.

Active sensors, on the contrary, emit a radiation and acquire a response to it, whether this response is reflected, refracted, scattered or emitted by the detected objects. Active sensors include, among others, LiDARs.

Relevant phenomena can be successfully investigated by the use of multispectral sensors. These are capable of recording images in the visible light and other spectral bands.

The same general concept is applied at greater level of detail, when hyperspectral sensors are used. These devices can discriminate a higher number of narrower spectral bands, allowing for more complex detection processes. This is a more complex acquisition technology, often requiring relatively larger platforms and specific processing procedures; nevertheless, their future large-scale application seems to be in the cards.

Kurtz and Buckley (2016) present a review of hyperspectral imaging in close-range applications. Depending on investigation needs, sensors can be either ground based or airborne.

The possibility of monitoring different bands of the electromagnetic spectrum at the same time allows for a deeper understanding of natural and anthropic phenomena. These may have important effects on ecosystems and human life, although they were difficult to detect in the past at a satisfactorily large scale.

In geographical studies, near-infrared (NIR) and thermal infrared (TIR) bands are particularly useful. The former is generally associated with remote sensing of vegetation. The latter can also be useful in this field, but it proves effective in the detection and evaluation of thermal anomalies caused by human activities.

If we consider visible light, near-infrared, and thermal infrared sensors mounted on small drones, the full range of opportunities becomes evident. In spite of the many possible technical solutions, some types of data acquisition and processing are in fact recurrent. Hence, they can be applied to several different fields for different purposes. For instance, we can consider that a wide set of thermal anomalies can be spotted by the same equipment. The system may then be used to spot animals in wildlife monitoring, to detect energy dispersions in buildings or industrial plants, or even to observe the discharge of pollutants into a freshwater reservoir.

Similarly, multispectral cameras can be used for mapping vegetation properties at a certain moment of the growing season. Whether the observed plants are trees in an urban green area, crops on a field, woodland plants, or even weeds on a large decaying building, this is still within the applicability of sensors. The difference is in the analytical and interpretive protocols.

In other words, the effectiveness of the same technical solutions may be recurrent in different fields; appropriate modular configurations can exploit this advantage. Hence, small drones can cover a wide range of geographically relevant observations by relatively minor adaptations of their systems and procedures.

3.2 From Data Acquisition to Interpretation

Remote sensors are implemented in several different types, models, and versions. Their specific features vary, obviously, based on the platform which is meant to transport them. Their working principles are relatively similar, though. In theory, it is possible to integrate and scale the data provided by homologous sensors regardless of them being mounted on a satellite, a manned aircraft, or a drone. This is not always obvious, as it is often necessary to perform adjustments and corrections.

A concept that remote sensing experts are familiar with is how resolution (whether spatial, radiometric, and temporal) affects the ability of a certain system to actually "see" a given phenomenon. The actual sampling capability of a system defines also, as a matter of fact, conceptual and epistemological boundaries. What appears to be the case in applying small UAVs to geographic observation is that platform flexibility adds to the sensors inherent performance and contributes to the overall result.

In some cases, UAV-based data can be evaluated manually and visually. Expert analysts examine data and are able to provide information about phenomena. Sometimes analysis is purely visual; sometimes it can rely on specific quantitative manipulation. Traditional aerial photograph interpretation is a good example of the former. The latter can be demonstrated by manual processing and analysis of aerial thermography.

The overall trend, however, is toward ensuring an increasingly faster and more efficient quantitative accuracy as a base for interpretation. Particularly, major developments involve automation of time-consuming, cumbersome, and error-prone operations.

An easy example, very familiar to small-UAVs users, is aerophotogrammetry. This century-old set of techniques is intended to allow for the tridimensional reconstruction of physical objects from bidimensional images. It is not thematically new. For more than a hundred years, it was successfully used for different types of cartography and visual rendering. Traditional photogrammetry, however, required expensive and complex metric equipment and the strict planning and execution of surveys. For this reasons, its use was limited to some applications, mostly at high technical level and for highly strategical, institutional, or administrative purposes.

Later, the general progress in computer graphics, computer vision, and related algorithms allowed for the implementation of photogrammetric software by different operation standards (e.g., SfM, or structure-from-motion detection). This revolutionized the sector by making the overall technique far more accessible. Aerophotogrammetry has become more flexible and effectively deployable in large-scale cartography (Saadatseresht et al. 2015). The popularity of the new systems is leading to an obvious commercial boom. The number and variety of uses is also increasing. Aerial imagery and photogrammetry are now very common, although quality and scientific level of results vary from case to case and remain heavily dependent on external factors such as flight conditions (Mesas-Carrascosa et al. 2016).

Feature extraction and object classification are also gaining importance.

From a conceptual point of view, automated feature extraction and classification allow to acquire information about the presence and nature of a certain geographical "object" in an individual image or series of images. The objects are defined within classes, created according to a certain ontology. Instances for each class are identified by different possible semantic approaches. Consequently, the possibility of successfully extracting, classifying, and interpreting the nature of the observed elements depends largely on the ontology (or groups of ontologies) which regulate the classification. The nature of input data and acquisition techniques, naturally, plays an important role.

In satellite and high-altitude airborne remote sensing, geographic phenomena are often detected at medium and small scale. In this type of images, the "pixel," i.e., the minimal spatial resolution unit, corresponds to an often non-negligible area on the ground. The distance between two adjacent pixel centers corresponds to the minimum "ground sampling distance" (GSD), i.e., to the minimum "size" on the ground that can be discriminated by the sensor. Any object whose dimensions are smaller than the GSD cannot be "seen" as such by the sensor.

Based on this limitation, in satellite and traditional airborne remote sensing, it was possible to discriminate only objects whose areal dimensions were relatively large. Although it was not a necessary implication in principle, this practical fact oriented interpretation protocols toward geographical objects whose extent was not minimal. As this was assumed to be unavoidable, reflection on themes and procedures to recognize specific phenomena ended up neglecting small objects.

Widespread interpretation models in the pre-UAV era were typically "pixel based." They were generally focused on the statistical analysis of quantitative values associated to individual pixels. Several digital classification methods were then developed based on this approach. Per-pixel analysis was a starting point from which several different classification models were developed. As sensors and acquisition techniques developed, their ability to spatially discriminate smaller objects increased. It therefore became possible to "see" objects which were previously excluded from the analyses. New wide thematic domains emerged at that point, and new conceptual paradigms for handling them became necessary (Castilla and Hay

2008; Blaschke et al. 2014). Attention then moved toward techniques and methods for extracting specific *objects*, rather than more general *surfaces*, from images.

The challenge was then to follow semantically developed criteria. The objects were in the geographical space, and so they had to be described based on their nature and specific locational context, as recognizable in the images.

This theoretical approach – from which sets of workflows are now being derived in various implementations – is known as OBIA (Object-Based Image Analysis). The specific geography-oriented OBIA is called GEOBIA (i.e., geographic OBIA) and is currently one of the most promising frontiers of UAV-based remote sensing. This approach is likely to produce remarkable developments in the future, in spite of the mixed results obtained so far.

As small drones are capable of taking data at very large scale, detailed object and context analysis becomes appropriate. Blaschke et al. (2014) indicate that GEOBIA assumes analytical paradigms which are strongly associated to conceptualizations from geographic information science. This is so true that GEOBIA can actually be considered a subset of GIS. It interprets geographical objects as scalable systems from spatial and functional points of view. As such, they can be inserted into hierarchies and be interpreted according to specific ontological categories (Lang et al. 2004).

3.3 Browsing Through the Experiences

This section contains a summary of typical small-UAV applications in geography and geography-related fields. Without any ambition of exhaustivity, we intend to provide an overview of some applications, highlighting several aspects that we consider important in each topic.

According to the epistemological canons of physical and human geography, the way we aggregated subjects in the following pages may sound heterodox. In fact, the thematic connection is given by the specific perspective that drones can provide. UAVs allow to observe physical spaces in terms of locations, geometries, volumes, radiations, and variations of the former through certain time intervals.

What comes after data *acquisition* and preliminary *processing* is geographic *interpretation*, and it falls within the domain of its respective thematic context. In our discussion, however, we found it more effective to consistently keep the typical perspective of UAVs on various phenomena. This means focusing, most of all, on how similar acquisition and processing methods pave the way to various different application fields and interpretation areas. A general remark must be added at this point concerning artificial space as it can be observed.

The study of reification is one of the most relevant and challenging fields in which drones can be used. Considering many case studies in the literature and news from the Web, it is clear that some applications are mature, while others are in development. Others appear to be in the status of demos and prototypes. In spite of this, we wish to propose a somewhat general overview.

First of all, drones can help researchers to understand geometries of built objects and to recognize material patterns, flows, and emissions. Full evaluation of the mechanisms underlying those phenomena, however, may require a high level of interpretation, often assisted by methods and techniques of geographic information science.

A typical application of small UAVs, either alone or integrated with airborne remote sensing, is certainly multiscale evaluation of anthropic actions. The term reification indicates the material shaping of spaces as actuated by human groups. By using the available resources, spaces are built, modified, removed, and organized. Such activities are expected to distinctively alter the environmental configuration. Change may involve geometries (e.g., by excavation, construction, artificial modification of natural shapes, etc.) or energy emission/reflection patterns (e.g., by increasing or decreasing albedo, by creating or removing thermal sources, by releasing gases, pollutants, humidity, etc.). In many cases, these anomalies are detectable in the visible light, thermal infrared, and near-infrared bands. Obviously, for detecting and identifying many other categories of phenomena, other traces must be considered. However, the three aforementioned sets of radiations are in the detection domain of standard multispectral cameras as well as in that of today's normal RGB, NIR, and thermal cameras. This means that many reification-related phenomena can be easily spotted and interpreted – either manually or automatically – by the use of (relatively) low-cost sensors.

This allows to assist in monitoring human activities and to support various forms of territorial management and planning.

3.3.1 *Reconstruction of Space and Mapping*

The potential of acquiring reliable and detailed bidimensional and tridimensional cartography is one of the main reasons for which small UAVs are so popular. Post-processing of data is usually necessary for integration with other sensing devices and GIS (Petras et al. 2016). Accuracy and precision are typical issues in any drone-based survey. Although often less accurate than traditional, top-level cartographic acquisition tools, UAVs provide good area coverage and overall performance. This can also be further fostered by appropriate path planning algorithms (Li et al. 2016; Nedjati et al. 2016). In most cases, results of UAV-based cartography are more than decent, frequently achieving decimeter and centimeter accuracy in high-profile protocols. Submeter accuracy may prove acceptable in other cases, depending both on scales and purpose. Mesas-Carrascosa et al. (2014) evaluated standard UAV-based acquisition techniques involving the use of ground control points (GCP) and aerial triangulation to assess the accuracy of ortophotos obtained from a multirotor UAV. The authors' goal was to evaluate the performance of drones based on official mapping agencies standards. In that experience, the 1:1,000 scale proved the most appropriate for orthophotos.

Higher accuracies can be achieved by adopting very close-range, large-scale acquisition protocols. This approach is typical, for instance, of architecture and archaeology.

A recognized advantage of drone-based cartography is that it can be built fast and with relatively modest organizational support. Both circumstances are very important if urgent mapping is needed for any reason (Choi et al. 2016; Fazeli et al. 2016; Mehrdad et al. 2016; Sakr et al. 2016). Relative ease in the image acquisition and cartographic workflow is also convenient when periodical monitoring and change detection are the goal (Al-Rawabdeh et al. 2016; Saur and Krüger 2016).

Advanced solutions are also being tested, aiming to enhance remote sensing performance by sophisticated configurations. UAV-borne laser scanners, in particular, are worth attention. The recent possibility of installing them on small drones boosted UAVs' performance in high accuracy documentation of physical spaces. Their use is still limited by their cost and overall complexity, but this technology is suitable for current specific applications, and it is also quite promising for future development (Pilarska et al. 2016).

High-resolution laser scanning can be integrated with hyperspectral imaging (Gallay et al. 2016). At present, however, the most widespread approaches involve more accessible hardware, such as non-metric cameras. When low-cost and/or off-the-shelf consumer-level products are used for drone-based reconstruction and mapping, several issues must be taken into account.

Much depends on both "internal" and "external" factors. The current robustness and effectiveness of image processing may be deceiving about its ability to overcome poor sensor quality and unfavorable environmental conditions.

Among internal factors which may degrade cartographic performance, three appear to be typical: camera quality, camera-platform compatibility, and acquisition procedures. External factors are sometimes difficult to properly budget, but they turn out to be relevant. According to Wierzbicki et al. (2015), for instance, bad weather may severely affect camera performance in terms of accuracy (up to an average 25% degradation, in the case of that research).

From relatively early in their appearance on the civilian market, small drones were fitted with GPS hardware, mostly for navigation purposes. These devices are generally non-differential receivers and their accuracy is modest. Some typical inaccuracies can be partially solved by the integration with other avionics. For instance, onboard barometer and IMUs could help in holding altitude within narrower error than that affecting satellite-based navigation systems (Mader et al. 2015). Still, the available positioning quality may be insufficient for large-scale mapping. At this stage in the evolution, onboard non-differential sensors are useful in obtaining an approximate geolocalization of data. In many cases, topographic precisions can be reached by correcting errors in the overall image coverage. This can be done by establishing a reference to ground control points (GCP). These can also be used for absolute georeferencing of the overall model/mosaic.

The actual position of GCPs on the ground can be measured by traditional total stations (TST) by DGPS and RTK. The appropriate distribution and localization of control points is often essential for a successful enhancement of cartographic accuracy (Lo Brutto et al. 2014; Tscharf et al. 2015; Long et al. 2016). It is therefore a standard practice, even though it turns out to be the most time-consuming and organizationally invasive part of the survey protocols. In order to avoid this inconvenient, several methods and techniques have been developed in recent years. Many address the problem by implementing direct georeferencing, by the application of higher accuracy positioning systems/methods of different types. Once the goal is reached, reference to GCP might become unnecessary in several applications (Chiang et al. 2012, 2015; Eling et al. 2015; Hamidi and Samadzadegan 2015; Fazeli et al. 2016; Hosseinpoor et al. 2016; Yeh et al. 2016).

The ability to automatically define the spatial position of a drone over terrain may involve both mapping workflows and automated navigation.

Some of the possible solutions consider vision-based simultaneous localization and mapping (SLAM) (Munguia et al. 2016), vision-based pose estimation and gravity vector measurement (Kniaz 2016), "image network generation of uncalibrated images with low-cost GPS data" (Huang et al. 2016), post-processing kinematic (PPK) (Frith 2017).

The advent of low-cost onboard RTK receivers on off-the-shelf UAV hardware has drastically reduced the overall effort required for enhancing spatial accuracy of drone-generated cartography.

On the image processing side of the workflow, one must emphasize the important developments obtained in recent years. New image elaboration algorithms on the one hand and more powerful implementations on the other hand allow to generate accurate models based on easily acquired data. Formerly invasive activities are being minimized. While photogrammetry develops as a primary sector for small drones, de facto standards seem to involve relatively low-cost hardware and software in professional applications (Lucieer et al. 2014; Smith et al. 2014; Balletti et al. 2015; Yanagi and Chikatsu 2015, 2016; Arseni et al. 2016; Tampubolon and Reinhardt 2016).

In general, large-scale mapping seems to be quite natural for small UAVs. Available tools appear to be well standardized, and more advanced ones are in further development.

3.3.2 Ambient and Atmospheric Observations

Atmospheric research and information are fields for UAVs in general and for medium and small UAVs in particular. If appropriate measurement techniques are developed, they can be of use toward understanding natural phenomena as well as effects caused by human presence.

Meteorologists are familiar with one of the most ancient types of UAV, i.e., the weather balloon. These small aerostats are commonly used as vectors for radio-probes. They collect data about atmospheric variables and downlink them to ground stations for status monitoring and forecast development. Going back to a previously described distinction, it would be possible to say that weather balloons are UAVs in the most basic sense of the concept: they are unmanned aerial vehicles, but since their flight cannot usually be controlled, they are not RPAS (Remotely Piloted Aerial Systems), unlike the other machines discussed so far.

Villa et al. (2016a) review the general characters of small and medium UAV technologies in atmospheric studies. Both fixed-wing and rotary wing platforms are used. Fixed wing provide a wider area coverage and generally have a faster cruise speed; this allows to often obtain a better representation of phenomena diversity over longer distances. Rotary wings can more flexibly change their trajectories to provide a more detailed description of conditions over and immediately around a location of interest. Their flying principles, however, force users to adopt measures so as to avoid perturbations (Neumann 2013; Roldán et al. 2015). As it was discussed by Villa et al. (2016b), appropriate adaptations are necessary when using multirotors for air quality and pollution detection. Though very popular on the market and highly maneuverable, these machines are typically affected by intense propwash and aerodynamic turbulence. Such phenomena must be properly budgeted in designing sensor and inlets installation, as airflow can be disrupted in proximity of those crucial components.

In general, small and medium UAV-indicated applications include studies on atmospheric composition, vertical aerosols profiling (Corrigan et al. 2007; Bates et al. 2013), pollution, and greenhouse gas emissions (Khan et al. 2012; Renard et al. 2016).

Magnetic phenomena at low altitude can be studied by the use of appropriately designed and built small and medium UAVs (Sterligov and Cherkasov 2016).

Ambient radioactivity can also be studied by fitting UAVs with adequate instrumentation (Martin et al. 2016).

3.3.3 Physical Landscapes: Landforms, Morphology, and Built Spaces

3.3.3.1 Landforms and Geomorphology

UASs are effective in image-based surface reconstruction. In geomorphometry, recent developments and challenges have been reviewed by Eltner et al. (2016). The authors considered publications in the following fields: soil science/erosion, volcanology, glaciology, mass movements, fluvial morphology, coastal morphology. These are particularly promising sectors for the technology (Fig. 3.1).

UAVs can be used in landform inspection and survey, especially when close-range observation is required. Such is the case for inaccessible/dangerous locations.

A popular use is in early warning inspection and periodical monitoring of different types of landslides (Lucieer et al. 2014; Turner et al. 2015; Davis 2016; Fernández et al. 2016; Peppa et al. 2016) or locally impervious geologic formations/surfaces. Giordan et al. (2014 and 2015) indicate a general workflow for multirotor-

Fig. 3.1 SfM-derived tridimensional model of the western section of Tarcal Valley, Tokaj Region, Hungary. The model was obtained from 124 images acquired by a custom 3.1 kgs hexacopter (GREAL 600R), mounting a Sony Alpha 5000, 20 Mpx mirrorless camera with a 16 mm lens. The total flight time was in the order of 10 min, and the total processing time was 8 h on a HP410 standard laptop with an AMD A10-9600P Radeon R-5 processor and 16Gb RAM. The area coverage was about 5 ha. Considering the complex geometry of the site and the high level of detail achieved, the convenience of using a UAV appears evident (Image: GREAL – European University of Rome)

based inspection of landslides and rockfall phenomena. Vrublová et al. (2015) evaluate the use of a fixed-wing UAV in the photogrammetric documentation of landslides and inaccessible spots of a quarry, comparing data acquisition techniques and evidencing strength and weaknesses of different approaches. Slope mapping is also a topic (Tahar 2015). In these mission profiles, drones are generally supposed to allow for the morphological recognition of outcrops, surfaces, slopes, and related components. Depending on the location and nature of phenomena of interest, work may be limited to basic visual observation, or span from expeditive geometric reconstruction to detailed morphological survey. Recent applications involve the use of GEOBIA classifications for discriminating objects and ongoing phenomena (Bertalan et al. 2016).

When UAV-based data are modest in terms of accuracy and precision, integration with different ground-based surveying systems and reference measures, or with other techniques, is appropriate. Multi-technique acquisition, however, may be recommended, in some cases, even for higher-profile equipment, so as to foster level of detail and/or accuracy. On the other hand, appropriate registration protocols shall be adopted to ensure reliability of information acquired from different sources.

Tong et al. (2015) consider the integration between TLS and drone photogrammetry (particularly to improve geo-positioning accuracy). It can be noticed that although more precise, ground-based acquisition systems prove generally better in terms of strict metric accuracy when placed in favorable locations, the fact that drones fly may prove useful for quicker and wider area description as it was shown by Dewez et al. (2016). These authors compared the performance of ground-based TLS and UAV in acquiring the geometry of a collapsing cliff, in relation to potentially exposed assets. Pro and cons of the two techniques (alternatively tested) in geomorphological mapping are also presented in Tilly et al. (2016). A comparison between UAV photogrammetry and airborne LiDAR is published in Caprioli et al. (2016).

Important geological features potentially observable by drones are traces of faults and fault planes (Amrullah et al. 2016).

Other phenomena involving landforms, such as glacier motions, calving dynamics, and various types of snow coverage, can be at least approximately mapped – and volumetrically and/or radiometrically estimated – through UAVs (Whitehead et al. 2013; Ryan et al. 2015; Vander Jagt et al. 2015; Bühler et al. 2016; De Michele et al. 2016; Harder et al. 2016; Immerzeel et al. 2017).

UAV application to the observation of soil erosion is documented, among others, by d'Oleire-Oltmanns et al. (2012) and Wang et al. (2016).

Remote or relatively inaccessible geographical areas (polar regions, deserts, etc.) can be partially surveyed by drones as they may also extend the observation capabilities of ground crews in the investigation of those remote areas and of environmental or anthropogenic phenomena in high detail and good temporal resolution (Turner et al. 2014; Bollard-Breen et al. 2015; Fraser et al. 2015).

The opportunity of using rotary wing or fixed-wing UAVs in these scenarios depends, obviously, on the extent and configuration of the areas of interest. In narrow spaces, multirotors are better suitable, and if required trajectories are complex, it might be advisable to rely on manual control, possibly assisted by ground proximity sensors for automatic separation from the obstacles. In many other cases, fixed wing UAVs are more efficient in terms of coverage.

3.3.3.2 Reificated Space: Buildings

Survey of buildings is appropriate to tackle engineering, architecture, and urbanization problems. Geography kicks in once the opportunity rises to assemble wider maps in which individual buildings are meaningfully integrated with information about space utilization, functional flows, social use, etc. Scale obviously matters as far as resolution and accuracy requirements are concerned.

Morgenthal and Hallermann (2014) present a detailed assessment of UAV-based visual inspection of structures. An interesting aspect of this work is that it highlights the importance of qualitative observation by small UAVs demonstrating the potential of this technique, often neglected by the literature. The authors focus on rotary wing platforms discussing technical aspects such as the effects of vehicle motion, wind, and turbulence.

In this type of application, the UAV must be able to maneuver so as to guarantee the best possible coverage and inspection of built surfaces. In general, therefore, rotary wing platforms are best suitable for close-range structural inspection (Otero 2016). Both manual and automated navigation planning and control are

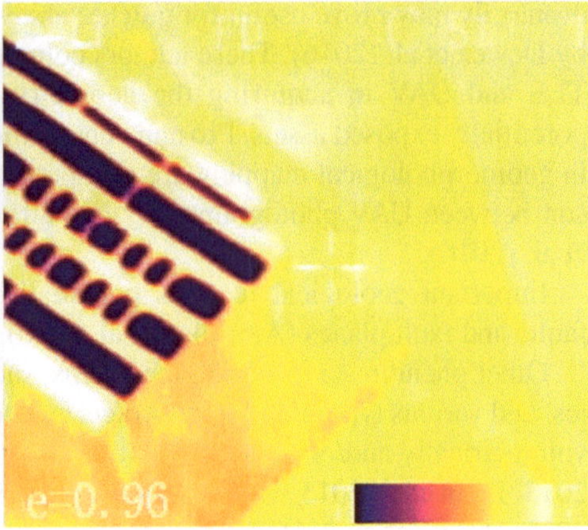

Figs. 3.2 and 3.3 By exploiting a known effect caused by wide difference in emissivity coefficients of materials, thermal images (*right*) associated with corresponding visible light ones (*left*) allow to discriminate polished metal elements (*darker* in the TIR) from composite elements on top of buildings, in spite of the similar color in the visible (Image: GREAL – European University of Rome)

commonly available and will be even more developed in the future. Image-based workflows for virtualization are common and may reach remarkable levels of technical refinement (Alidoost and Arefi 2015). Laser scanners are also used, for their generally better measurement performance as long as ranging can be performed at appropriate distance and angles (see Chap. 5). Frequently, integration of airborne and ground-based photogrammetry and laser scanning is indicated toward a more detailed and accurate result (Maiellaro et al. 2015; Zarnowski et al. 2015).

A time- and resource-consuming component of the aforementioned techniques is the establishment of control points. Such work remains nevertheless necessary due to the relatively small size of many features and their very complex geometries.

Direct georeferencing is now taking over as technology progresses. It drastically reduces time and effort toward a reliable spatial orientation.

It is worth noting that building inspection goes beyond mere visual documentation or more or less sophisticated geometric reconstruction. Since buildings are inherently complex objects in which several phenomena are to be monitored, specific types of observations are performed by the use of different sensors. Thermography is deployed to monitor thermal sources and their distribution within a given building or its immediate surroundings, along with the identification of anomalies related to malfunctions or construction issues. Also, thermographical images can be used as sources for recognizing other types of phenomena (Figs. 3.2, 3.3, and 3.4).

Near-infrared images can be useful for the detection of living vegetation on architectural elements, as this might be difficult to detect by RGB and thermal cameras only. The joint use of multispectral images could turn particularly useful if associated with building geometrical reconstruction. By adopting such an integrated

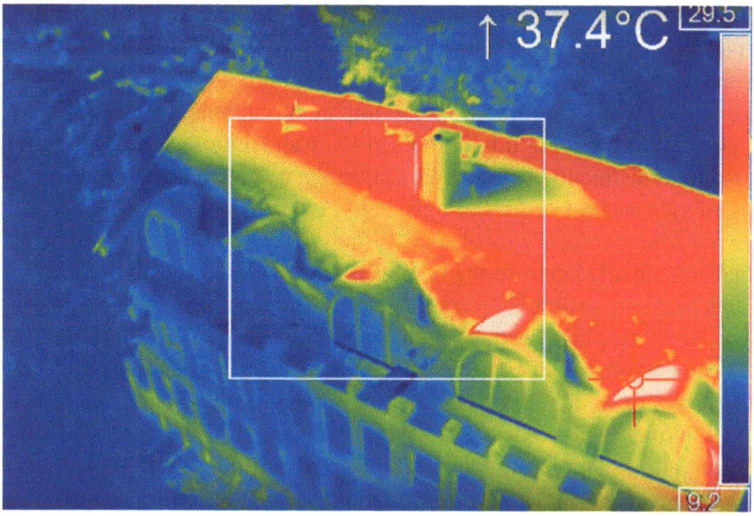

Fig. 3.4 Aerial thermal image of the surface of a building heated by sun exposure in daylight conditions. The section of the roof which is more directly hit by sunlight is warmer and releases higher thermal radiation. A tree in the foreground (on the *left*) shades part of the façade, whose temperature and emitted radiation are therefore lower. Although very accurate thermographic measurements require specific ground-truth and sampling procedures, drones can provide effective overviews on general thermal trends and anomalies, either natural or artificial. This picture is a screenshot from a video recorded by a small UAV equipped with an Optris PI450 thermal camera (Image: Process Parameters Ltd., www.processparameters.co.uk)

approach, ongoing phenomena can be mapped for easier data analysis and manipulation (Mader et al. 2016).

Among non-image-based techniques, Na and Baek (2016) envision the possibility of using small UAVs for monitoring building structural health by direct measurements through vibration-based methods with sensing devices installed on board low-cost multirotors.

3.3.3.3 Reificated Space: Anthropization in Wider Areas

The observation of human presence and activity in a given geographical area goes beyond individual building inspection. It involves techniques to acquire data and perform GIS-based analyses, with favorable opportunities in urban areas when accessible.

Observation of urban areas is among the most elementary type of activity a small drone can perform. At the same time, though, it is one of the most powerful in terms of knowledge that can be acquired and one of the most critical in terms of juridical issues and social perceptions. Civil protection and law enforcement could obviously take major advantage of the observational capabilities of small drones (Idachaba and Oni 2015), and several police services in various parts of the world are getting equipped with small drones, but a complete discussion of this topic would be out of the scope of this work.

One of the most natural kinds of geographic observation over reificated areas is about general land use. For this, both fixed-wing and rotary wing UAVs are suitable, with the usual distinction between area coverage and local maneuverability.

Prevailingly natural or prevailingly built contexts can effectively be examined, but the fundamental capability of small drones to fly low and get close to the targets is valuable in detailed monitoring of constructed bodies.

In this field too, visible light sensors are most effective in morphological reconstruction and mapping; current available tools are useful in monitoring change detection and evolution of construction and building activities (Unger et al. 2014).

This is the case, for instance, for quarries and mines, either active or abandoned. If active, their observation is important for cultivation-related economic, ecological, and functional aspects; if no longer active, then monitoring is appropriate as they are potential hazard or requalification sites (Haas et al. 2016; Raeva et al. 2016).

Like in other fields, developments in this particular area are pointing to progressive integration of acquisition and classification protocols. It should be noted, though, that in several practical cases, a generic geolocation of data or qualitative information might be perfectly useful.

In our review of both the scientific and "gray" literature, we have found several examples of observation and interpretation workflows based on different theoretical, methodological, and operational approaches.

These span from classification of elements in informal settlements (Gevaert et al. 2016), to accuracy measurements in earthworks (Daakir et al. 2015), to landfill management (Long and Clarck 2014). Straková et al. (2015) present UAV-based visible light and thermal monitoring of heaps.

Other cases involve OBIA analysis for extraction and classification of urban, composite, or rural areas (Sari and Kushardono 2015; Yu et al. 2016), integration of UAV data into multi-criteria GIS analyses for spatial planning (Zawieska et al. 2016), or geodesign platforms (Anca et al. 2016).

Other applications involve area mapping, site or infrastructure inspection by simple visual observation, metric acquisition, or by development or application of recognition/classification algorithms (Laliberte et al. 2010; Máthé and Buşoniu 2015; Wilner 2015; Suziedelyte Visockiene et al. 2016; Yeh et al. 2016). A potentially important sector is thermographic mapping of settlements. Since use/abandonment and energetic efficiency of buildings are becoming important factors in policy making, territorial management, and law enforcement, the availability of UAV-borne thermal cameras may appropriately respond to this type of surveying needs. TIR observation of settlements, in this sense, can be limited to qualitative reconnaissance, or be extended into quantitative and geospatial description by appropriate classification procedures, GEOBIA in the first place (Hay et al. 2010).

3.3.3.4 Reificated Space: Infrastructures and Flows

Life in any region depends on flows of persons, materials, energy, and communication across spaces and territorial networks. Nodes in such networks include residential and productive sites and several types of infrastructures. Flows along the lines connecting these nodes allow for territorial systems to live and operate according to purposes, cycles, and demands of economic, social, and cultural life. The material effects of these flows can be monitored and so yield geographically relevant information. Small UAVs are one of the possible tools for observing some of these effects.

Massive movement of persons and goods takes place by the use of standard transportation devices and vehicles. Their concentration in certain areas (e.g., industrial complexes, loading and parking areas, stations and ports, etc.) at certain times, storage of materials and containers, and circulation of specific vehicles along the road network can be statistically relevant indicators of industrial or commercial activities. Other types of material transfer, if appropriately detected and interpreted, can be relevant in understanding flows.

Even detectable externalities of otherwise undetectable flows can serve as a valuable indicator.

Drone-based systems can be used for qualitative monitoring of several types of activity.

Some UAS are not only capable of monitoring vehicle activity but also crowd distribution. Even pedestrians can potentially be tracked and globally monitored or individually followed. As applications of this kind develop in the future, the nature of their deployment and limitations to their use will likely be based not on technical considerations but rather on juridical and ethical criteria.

The future ability to monitor traffic flows by UAV platforms – especially since the complexity seems more in the recognition process than in the acquisition technology per se – may provide a very strong support to urban studies as well as to territorial planning (Rodríguez Canosa et al. 2012; Yuan et al. 2015; Ma et al. 2016; Xu et al. 2016a, c).

3.3.4 Rivers, Streams, Wetlands, and Water Management

Rivers, streams, and inland waters are important observation targets. This field demonstrates, particularly, the effectiveness of the large-scale and close-range philosophy embodied by small UAVs. It is now possible to investigate river stages (Niedzielski et al. 2016), riverbeds, and shores, including hydromorphological features (Rivas-Casado et al. 2015). Several phenomena can be studied with regard to water bodies and their interfaces. Among those, riparian vegetation conditions (e.g., Dunford et al. 2009; Husson et al. 2014), shore forms, and shallow water environments are worth mentioning. Visual observation of UAV-based images is useful, although research and work on this type of environment are likely to dramatically develop in the future. Developments may include integration of satellite, airborne, and UAV-based surveys and already explored OBIA/GEOBIA classification methods (Johansen et al. 2011; Dronova 2015; Demarchi et al. 2016). Different sensors make it possible to perform more sophisticated kinds of survey, such as double-medium profiling. This is the case of topobathymetric surveys in which morphology above and under water can be quantitatively evaluated by laser (Mandlburger et al. 2016) or by aerial photography. Both LiDAR and photogrammetric mapping require relatively low water turbidity (Flener et al. 2013). The combination of TLS and UAV photography, if appropriate ground truth is guaranteed, makes change detection of river morphology possible, over a given period of time.

Recognition of contamination in water bodies is always useful. In principle, several different techniques can be deployed for this purpose (direct sampling, detection of biological "markers," etc.). An easily deployable technique is aerial thermography, anytime the inmission of pollutants or foreign substances determines variations in the thermal condition of the liquid surface (Lega and Napoli 2010).

Inland waters and rivers are often widely connected to populated spaces, and they can be more or less heavily shaped by artificial measures. Artificially regulated waterflows are tightly related to territorialization in general and are major geographical factors, due to the complex dynamics they can trigger.

UAV-based observation and related GIS analysis are therefore increasingly oriented to studying these aspects.

Applications span from maintenance-oriented prospections (Kubota and Kawai, 2016) to identification of fluvial hydromorphological and morphodynamic configurations (Miřijovský and Langhammer 2015; Rivas-Casado et al. 2015, 2016), to disaster assessment and measurements (Perks et al. 2016), to studies about runoff phenomena. The latter appear to be critical insofar as they can affect sensitive inhabited spaces (Barreiro et al. 2014; Tokarczyk et al. 2015) (Figs. 3.5, 3.6, and 3.7).

Complex phenomena involving water accumulation and circulation on land surfaces can have important consequences in causing environmental degradation. The possibility of performing very high-resolution measurements and modeling of surfaces is important toward monitoring and action planning. Capolupo et al. (2014), for example, present an application of drone survey and hydrological models for detecting polluted areas at a large scale; Husson et al. (2014) present a study for mapping riparian vegetation species, biomass, and related metal contents.

3.3.5 Coastal Environments and Coastal Settlements

Coastline monitoring and feature identification in those areas are another trendy field for drones. Works involve cartographic processing, segmentation, and classification protocols by visible light and multispectral image acquisition (Ballari et al. 2016; Papakonstantinou et al. 2016). Natural coast and beach environments are subject to transformation due to geomorphological characters, tides, waves, and winds and other local conditions (Frith 2017). Observation of environmental dynamics is often particularly valuable, as coastal areas are important zones for human settlement, economic activities, and circulation of people and goods (Scarelli et al. 2016).

Figs. 3.5, 3.6, and 3.7 The safety of a road could be jeopardized by water runoff. A simulation is obtained by integrating UAV-based SfM 3D rendering of terrain (Fig. 3.5 *left*) with hydraulic modeling algorithms. An open-source software (DualSPHysics), based on the smoothed-particle hydrodynamics model, allows to represent water motion on the slope toward the road in the absence of any ditch (Fig. 3.6 *right, above*) or in the presence of a 0.8 m ditch in the area immediately uphill from the tourniquet (Fig. 3.7 *right, below*). This operationally simple and affordable analysis workflow has an evident value in fostering infrastructure safety (Image: Barreiro et al. 2014, see also www.dual.sphysics.org)

Proper and consistent monitoring over given periods of time can help in developing better knowledge of evolutionary mechanisms in these particular and sensitive contexts (Čermákova et al. 2016; Jaud et al. 2016; Lu 2016; Yoo and Oh 2016). Attention should be devoted in ensuring that UAV-based observation – especially when it is cartography oriented – is reliable and reaches adequate precision; otherwise, the monitoring would prove ineffective (Long et al. 2016).

Along coastlines, drones can be used for different types of qualitative and quantitative monitoring. Sophisticated applications include bathymetric surveys by airborne LiDAR (McLean 2015). However, bathymetric investigation by more affordable two-medium photogrammetry is also documented (Ye et al. 2016). Thermographic detection of pollution and perturbations can be conducted at different scales by the joint use of manned and unmanned platforms (Lega and Napoli 2010; Lega et al. 2012). Brouwer et al. (2015) propose an application of low-cost multirotor UAVs for surfzone monitoring at high temporal resolution.

3.3.6 Vegetation: Agriculture, Woodlands, and Forestry

UAVs perform generally well in vegetation monitoring, either in agriculture or forest studies.

Herbaceous vegetation can be monitored by UAVs (Clements et al. 2014; van Iersel et al. 2016). In general, Observation of vegetated areas covers many aspects. It spans from evaluation of plant general conditions in the same site, water stress, differences in growth and areal cover, etc. (Zarco-Tejada et al. 2011; Baluja et al. 2012; Candiago et al. 2015; Hoffman et al. 2016; Luna and Lobo 2016; Schirrmann et al. 2016).

Present-day sensor options include RGB; near-infrared, multispectral, and hyperspectral cameras and spectrometers; as well as other environmental measurement instrumentation. The separate or combined use of these systems allows to derive complex information from vegetation (Figs. 3.8 and 3.9).

Figs. 3.8 and 3.9 Samples of UAV-based orthomosaics for agricultural research. The image above (Fig. 3.8) is a raw multispectral image taken in the R-G-NIR bands (healthy vegetation in turquoise); the image below (Fig. 3.9) shows calculated NDVI values. The onboard sensor, a Canon A495 camera, was modified to record NIR values by removing the original NIR filter and was replaced by a Kodak Wratten 25 type filter (Image: János Mészáros)

As far as remote sensors are concerned, typically considered aspects are plant morphology and geometry on the one hand, and spectral signatures on the other hand, since the latter can give specific indications about biological conditions. Techniques and systems are to be developed accordingly (Honkavaara et al. 2013, 2016; von Bueren et al. 2014, 2015; Mesas-Carrascosa et al. 2015; Bareth et al. 2016).

A detailed review of tools and applications in vegetation studies is presented by Salamí et al. (2014) and gives a feeling of the work done by researchers and operators. The authors reviewed about 40 different experiments, conducted in 15 countries on 18 types of vegetation monitoring, by the use of 32 different UAV models. Urban vegetation, usually intermingled with more or less heavily built spaces, can also be studied taking advantage of small UAV flexibility.

A recurrently relevant parameter to be observed, both in agriculture and forestry, is the ground-canopy (i.e., plant height) distance. It is used as a reference for growth/health assessment, biomass estimations, etc. It involves discriminating, in detail, the vertical separation of the ground from the top of the plants on an areal basis. This is often achieved by considering – and measuring by various differential techniques – the shape of the ground in the area. It is represented by the DEM, or digital elevation model, i.e., what the terrain would be without any vegetation. It is a distinct surface from the DSM (digital surface model, i.e., the surface of the area at the moment of survey, including vegetation coverage). How these models are actually elaborated presents different qualities and technical aspects depending on the acquisition technique; laser scanning (both terrestrial and airborne) is used. UAV-based photogrammetric point clouds are also chosen due to their decent accuracy and overall accessibility in general plantation management (Bendig et al. 2014; Kattenborn et al. 2014).

Small UAVs are obviously becoming popular in agriculture, either for research or application purposes. New techniques are being experimented, and a wider presence of this technology in countrysides all over the world is foreseeable. Particular advantages for widespread use of UAVs are that agricultural areas are generally depopulated; hence small drones are seldom subject to restrictions; this allows more freedom in choosing configurations. Additionally, in comparison to other scenarios, farming areas are contexts where flying a drone is relatively easy from an operational point of view. Spatial patterns tend to be regular, so flight planning, ground-truth, or GCPs surveys present less difficulties than corresponding work in other types of site.

As of this writing, the agricultural sector appears as one of the future chief users of small UAVs. Indeed the new technology fits very well with the very large-scale approach followed by precision farming (Stafford 2000; Lelong et al. 2008; Zhang and Kovacs 2012). Michael Mazur (2016) has identified six application sectors in which drones are going to "revolutionize agriculture": soil and field analysis, planting, crop spraying, crop monitoring, irrigation planning, and health assessment. Although all of these functions are already in the domain of technical feasibility, some aspects of them are more developed than others. Yet, UAV technology is certainly going to enable activities in the entire aforementioned possibility range, and probably more.

As far as agriculture is concerned, drones are useful in clarifying vegetation trends: indeed, correspondence between what is observed and the underlying biological phenomena is rooted in long scientific and practical experiences; drones add the advantage that they can easily cover large areas, in great detail.

UAVs are actually capable to provide users with quicker, more detailed, and sufficiently reliable knowledge about the phenomena of interest. This has a direct positive impact on financial aspects and profit.

Given the favorable technical and juridical conditions and the strong interest for this kind of observation, UAV applications in agriculture are increasingly developing on a wide range of techniques. From affordable and consolidated SfM-based procedures, to more innovative acquisition and elaboration approaches, to sophisticated multispectral and hyperspectral analyses, up to integration with other forms of remote sensing, plenty of data can be obtained. Current work covers various sectors: soil and crop mapping, phenotyping, biomass estimation, canopy development, crop health and growth monitoring, plant traits, forage availability, weed management, and yield prediction (Mathews and Jensen 2013; Torres-Sánchez et al. 2013; Bendig et al. 2014; Clements et al. 2014; Borra-Serrano et al. 2015; Burgos et al. 2015; Capolupo et al. 2015; Peña et al. 2015; Baofeng et al. 2016; Di Gennaro et al. 2016; Holman et al. 2016; Lukas et al. 2016; Müllerová et al. 2016; Nebiker et al. 2016; Possoch et al. 2016; Sona et al. 2016; Shi et al. 2016).

Common features of UAV-based observation in agriculture involve a combination of remote sensing (either geometric-morphological and/or spectral) and appropriately periodized monitoring.

Other uses of UAVs in agriculture include substance spraying at specific locations on the fields. This indicates a technological trend, despite the often limited payload capabilities of small drones. Yet, research is developing in this sector (Faiçal et al. 2014; Pan et al. 2016). Among other innovative applications of small drones to agriculture, one should mention research developments regarding the use of UAVs in greenhouses as part of wireless sensor networks toward monitoring of several environmental variables (Roldán et al. 2015).

Forest, woodland, and, more generally, tree monitoring by small UAVs is relevant to both scientific research and, beyond that, to resource management (Wing et al. 2013). Phenomena are generally studied based on the development of medium or high accuracy cartography, for which several techniques can be integrated. Observational aspects of interest are tree structure and cover – OBIA is already applied in traditional remote sensing (Abbas et al. 2010; Delaplace et al. 2010) – growth development (Karpina et al. 2016), gap patterns (Getzin et al. 2014), and vegetation health aspects on either areal or per-plant basis (Zmarz 2014; Czapski et al. 2015; Smigaj et al. 2015). Inoue et al. (2014) present the use of UAV in surveying distribution of fallen trees in deciduous broadleaved forests as relevant for biodiversity research. Feng et al. (2015) discuss a workflow for classifying urban vegetation.

Studies may focus on disturbances and pest and disease infestations (Näsi et al. 2015; Minařík and Langhammer 2016; Yuan and Hu 2016). Sensors may include multispectral and hyperspectral devices, but also separate visible light, near-infrared, and thermal infrared cameras. The use of SfM-derived point clouds – possibly integrating ground-based and airborne acquisition – appears to be suitable for forest structure analysis (Mikita et al. 2016, Honkavaara et al. 2016), as the technique is progressively standardized and enhanced in its procedural aspects. Particularly when acquisition is done by UAV only, however, "altitude, overlap, and weather" conditions remarkably affect the accuracy of photogrammetry-generated point clouds (Dandois et al. 2015). Airborne laser scanning (ALS) is also considered, as miniaturization allows to install sensors on small UAVs. This type of acquisition could be more suitable when highly accurate geometric measurements are required, e.g., in terms of vertical distribution of vegetation (Wallace et al. 2016).

In many kinds of vegetation studies and surveys, involving the use of small UAVs, ground surveys are performed for validation and to obtain additional data.

3.3.7 Wildlife and Livestock

The use of flying platforms to monitor animals is common in both livestock and wildlife management.

Relatively wide area coverage is important in this type of application, which can effectively be performed by visible light (VIS) and thermal infrared sensors (Chrétien et al. 2015). These authors also refer to a multi-criteria image analysis (MOBIA) method which indicates the feasibility of identifying species through automated recognition by cross-reference between VIS and TIR.

Observation by VIS images (either still or video) is a common practice for both scientific research, divulgation, and environmental protection. Joint or alternate use of VIS and TIR sensors is appropriate so as to reliably spot animals in day/night conditions. In normal conditions – especially by night – animals tend to have a higher temperature than other objects around them. This makes it possible to spot their presence by the use of drones fitted with thermal cameras once correct identification procedures are defined. If the purpose of the observation is to "see" individuals or monitor their transfer across a certain area, then the drone can provide a solution; counting the animals may be far more challenging, as their ability to move relatively fast may bring to false positives or negatives as animals might move while the drone is inspecting the area (Prieto et al. 2014).

Observation of wildlife in its habitats is a major opportunity for environmental and natural sciences. Linchant et al. (2015) review a series of UAV applications to wildlife monitoring, presenting strengths and weaknesses of different systems and approaches. According to the authors, fixed-wing UAVs are generally the most appropriate tool for this application, since flight endurance is a major requirement and fixed-wing UAVs provide wider coverage than rotary wings. The authors also summarize the results of several publications reporting on UAV performance in effectively recognizing animal species and features in different contexts.

Several aspects can be more or less successfully monitored: the approximate number of individuals in a given area and their general conditions; even behaviors are easier to observe than by the use of traditional manned aircraft. Indeed, small drones, besides being cheaper, could be perceived as less invasive by animals (Ogden 2013; Bourne 2016).

Opportunities are now emerging to document wildlife by appropriate expeditive cartography. Qiu (2014) listed the applications in which UAVs could significantly contribute to wildlife monitoring and protection: (1) fighting wildlife crime, (2) close-range documentation of otherwise inaccessible or unapproachable animals, (3) population count, (4) wider area animal-in-habitat monitoring, and (5) wildlife protection space and infrastructure monitoring.

Clark Howard (2016) presents a mapping project for monitoring the community of sharks of an atoll in the Seychelles archipelago, obtained by a fixed-wing small UAVs fitted with high-resolution VIS camera.

Promising results are being obtained by several approaches to automated classification/recognition of individual animals (Chrétien et al. 2015; Zmarz et al. 2015; Gonzalez et al. 2016).

3.3.8 Disasters

Small UAVs are very effective search and observation tools in case of catastrophic events. They are useful in quick response, relief operations, and scientific study of disaster dynamics.

Drones can be used to observe many different kinds of impeding and unfolding disasters: floods, earthquakes, avalanches, pollutant spills, dam and industrial plant failures, and fires. The possibility of achieving rapid mapping is important to emergency operators in order to get adequate situational awareness about the evolving phenomena (D'Alessandro 2016). A discussion of aspects involving deployment of UAVs in disaster scenarios is presented in Tanzi et al. (2016).

In general, the choice for one type of platform instead of another reflects the previously stated considerations about the operational scenarios: rotary wing platforms – especially multirotors – are very useful in applications that require local-scale precise maneuvering and hovering; such is the case, for instance, when per-building inspection is appropriate. Fixed-wing UAVs are capable of longer range and, ceteris paribus, wider area coverage. Boccardo et al. (2015) explicitly indicate rotary wing platforms as the most appropriate for interventions in case of industrial accidents (including indoor prospections) and landslides; fixed-wing platforms in case of floods and wildfires.

Purely inspective, qualitative observation and rapid cartographic rendering in the visible light is generally effective for well-descriptive views of the overall situation (Bendea et al. 2008; Constantinescu 2013; Christensen 2015; Perks et al. 2016). Quantitative and more analytical surveys can be performed so as to acquire more data than by means of traditional reconnaissance. It is worth noting that a crucial aspect of UAV operation in disaster scenarios is area coverage. This is necessary both when the drone is used for observation and when it is used for other purposes.

Popescu et al. (2016) refer to trajectory optimization for aerial coverage; Sánchez-García et al. (2016) propose a strategy for applying artificial intelligence algorithms to UAV navigation and self-deployment in disaster scenarios. The specific purpose of the latter authors is to use UAVs as communication relay, even though the principle would be effective for reconnaissance as well.

Focusing on quick response, specific strengths of small drones in this particular kind of application are easy deployability and low absolute organizational overhead. Obviously, limited range may be an issue, but in most emergency cases, drones are supposed to complement other platforms already providing for wider coverage. On the other hand, the capability of small UAVs to get very close to the ground is essential in spotting critical situations which might remain unnoticed from classical manned platforms such as helicopters or airplanes. This capability is also relevant for material inspection and damage assessment at large scales. It should be noted that in first response during emergencies, high resolution of images might be secondary, in some cases, to robust and lightweight data transfer. This may lead to specific optimization solutions (Musić et al. 2016). Semiautomated target recognition can provide an important advantage too (Sun et al. 2016), but in this case the capability of systems to actually spot a certain element remains heavily dependent on image quality.

In all cases, different types of sensors and protocols may be used to facilitate location and identification of targets: from plain digital imagery to thermography, up to signal detection (Moon et al. 2016), etc.

An often neglected opportunity given by multirotors is that this type of platform can actually fly indoors; provided that appropriate measures are taken to prevent damage to the drone and interruption of data link, small UAVs become effective inspection tools inside large damaged buildings (Casagrande et al. 2015).

As far as detailed investigation of disaster effects is concerned, a clear opportunity is the possibility of integrating small UAVs with other data acquisition technologies in the monitoring of damaged structures (Balletti et al. 2014; Achille et al. 2015), or detection of landscapes change after disasters (Izumida et al. 2017). It is remarkable, particularly, that several types of classification and also OBIA begin to be applied in the semiautomated analysis of damage to buildings, so as to foster high-detail classification of disaster effects at large scales (Fernández Galarreta et al. 2015; Chen et al. 2016).

Another obvious kind of threat in which drones are being used is wildfire; applications are obvious in planning (Alonso-Benito et al. 2016), early warning, up to scenario monitoring. Although range and payload considerations suggest that many activities in a wildfire-related emergency can be developed by the use of larger and manned platforms, still it is worth noting that drones can be a useful complement as they can be more massively deployed on the scene and they can often be flown closer to the danger zones.

Li et al. (2012) propose an integrated workflow for using UAVs, flood simulation, and GIS processing for emergency evacuation planning in possible dam-break scenario.

3.4 A (Brief) Journey Through Human Time and Space

So far we considered the potential of small drones in observing physical spaces and variables. In this section we would like to reflect on their possible use for the purposes and methods of historical, cultural, and social geographies.

3.4.1 Archaeology: Spaces Through Time

As it was previously stated, landscape – as an economic, social, and cultural build of space – has a texture which evolves in time.

A specialized science developed around this concept is aerial archaeology, now a well-established and mature activity field in terms of epistemology and methods. It investigates past reification by examining its residual traces in current landscape.

Born in the early twentieth century and developed for over a hundred years by manned aircraft, aerial archaeology looks for variations in soil color (or multispectral signatures), vegetation growth, and small alterations in terrain shapes. Proper analysis of these phenomena may lead to discovering or better describing ancient buildings and settlements.

Small drones and related technologies are being successfully deployed in this kind of research. As a matter of fact, they significantly fostered some aspects of its overall scientific outreach. The possibility of flying so close to the ground and so close to objects made it possible to adopt different perspectives, levels of detail, and operational profiles. An intriguing and innovative – though somewhat extreme – example is presented by Louiset et al. (2016) who demonstrate the potential of a minimal-sized multirotor to allow for three-dimensional documentation of individual objects. All of this goes far beyond the applicability of traditional manned flight. Small UAVs are indeed used today for acquiring data within a specific excavation (Eisenbeiss and Sauerbier 2011; Ortiz Coder 2013; Malinverni et al. 2016). At the same time, the efficiency of UAV-based survey allows for comprehensive modeling and mapping of relatively wide archaeological sites and areas (Uribe et al. 2015; Ostrowski and Hanus 2016). Techniques span from application of simple aerial views to aerophotogrammetry and multispectral mapping and to the use of laser scanners in various forms of integration. The progress of video/photo capture is making UAV-based documentation an everyday standard practice, and it is pushing toward new developments. Xu et al. (2014, 2016b) proposed the integration of TLS data and UAV-acquired imagery, respectively.

It should be noted that, along with satellite and traditional airborne observation, UAVs are useful for spotting or identifying previously unknown or scarcely documented archaeological sites and landscapes (Casagrande and Salvatori 2011; Parcak and Tuttle 2016; Pavelka et al. 2016).

This fact, along with faster data processing techniques, provides researchers and operators with a new capability of performing more work in less time and effort.

A well-developed modus operandi involving the study of cultural heritage calls for the integrated use of small UAVs with other ground-based technologies for mapping, inspecting and surveying ancient buildings and their surroundings (Salonia et al. 2011; Ballarin et al. 2013; Patias et al. 2013; Chiarini et al. 2014; Lo Brutto et al. 2014; Meschini et al. 2014; Muñoz-Nieto et al. 2014; Russo and Manferdini 2014; Leucci et al. 2015; Sun and Cao 2015; Reiss et al. 2016).

Although the techniques and remote sensing technologies in this fields are not very different from other used in observing normal, present-day buildings, historic ones have specific relevance due to their long time of existence.

A deep knowledge of their structural and material conditions involve understanding decay of certain components, detection and investigation of construction/restoration phases, etc.

The importance of this kind of work is at the same time culturally and materially relevant, as it provides information about past reification insofar as it is integrated in currently existing systems and spaces.

Important perspectives for the representation of historical spaces come from new technical opportunities of producing multimedia documentation and virtualization of cultural heritage assets. Accessible workflows from data acquisition, to elaboration of virtual models and presentation of those, allows to envision new techniques – along with traditional ones – for spreading awareness and knowledge about culturally or socially relevant places (Figs. 3.10 and 3.11).

Fig. 3.10 Aerial view of an area of the ancient Roman City of Aquinum (Frosinone, Italy). Traces of disappeared buildings can be recognized in the marks visible in the grassy area on the left of the image (Image: LABTAF, University of Salento)

Fig. 3.11 Nadiral fish-eye image of the Otricoli archaeological area. This type of view, which can be used as input for cartographic purposes, is also effective toward a better understanding of the diachronical relations among current landscape elements, visible ruins, and possible traces of other material remainings beneath the surface (Image by FlyToDiscover, www.flytodiscover.it)

3.4.2 Historical and Cultural Geography: Places Through Time

Are drones able to tell us something about cultural space that has not been said by other acquisition technologies? An answer is not obvious.

Drones can view spaces from different angles and acquire material for knowledge in a more flexible and – so to speak – "agile" manner than other systems and tools.

However, drones should not be overestimated: their effectiveness is always dependent on the users' ability to conceptualize image and data acquisition.

Drones can only "see" material elements: therefore, they can help in revealing aspects of cultural landscape only as long as "markers" of cultural meaning are visible in the scene. These markers can be still and permanent objects (e.g., buildings, signs, monuments), or they can be mobile and temporary (e.g., clothes, gestures, expressions). From this point of view, other visual acquisition devices, such as photocameras, smartphones etc., can provide the same types of information.

Cultural landscape, however, is not just manifested by what the scene *contains*; rather, it can also be described, in some ways, by how the scene is *taken* (Bignante 2010). The drone itself may become a tool of expression and, just like photo and video cameras, it may convey the "views" chosen and created by a certain social and cultural identity. The difference with traditional cameras is that the drone may add some degree of freedom to how these "views" are created.

Small drones, which are potential tools for participative observation in science (Grainger 2017), can also contribute to expressing geosocial and geocultural information.

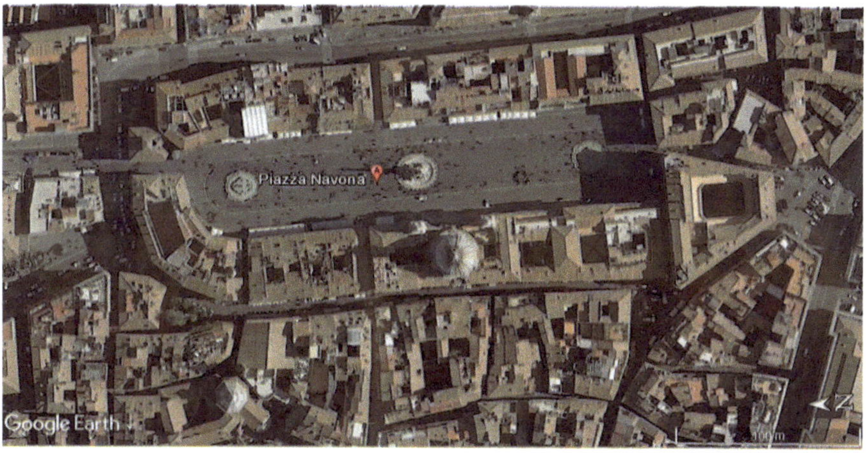

Fig. 3.12 Piazza Navona, Rome, Italy, is the result of the spatial "fossilization" of the ancient Roman stadium inaugurated by Emperor Domitian in the first century C.E. The currently existing buildings are mostly based on Roman foundations, structures, and walls of the ancient *stadium*. The square, a world-renowned landmark of Italian art and history, has retained a distinctive circus-like shape and a social value as a major meeting and leisure place

Since cultural geography is often deeply intermingled with the history of places, such historical dimension is very important and should always be kept in the researcher's mind.

In studying the evolution of places through time, material components of each phase could be aggregated in a set of diachronical layers. Elements in each layer can be erased as territorial evolution proceeds, or they can survive. Some kind of "footprint" of previous stages, at that point, may live into subsequent ones.

This appears as a persistence of spaces which remain physically recognizable through the evolutionary process. Parts of ancient configurations may be retained; others are changed, but new buildings keep some degree of resemblance to the previous ones. Even the social role of those spaces may evolve somewhat correspondingly.

As it happens in the fossilization of organisms, the inner structure of spaces may be progressively reconstituted. This occurs when initial elements are replaced, "molecule after molecule," by new components. The result is the transformation of previously existing geographical objects into new geographical objects with different functions and purposes. Yet, a physical, perceptive, and ontological continuity stands through chronological phases (Fig. 3.12).

"Spatial fossilization" is a complex phenomenon. It is relatively common in areas where human settlements date back to many centuries ago. In fact, it can be more or less evident depending on how much need or interest a certain culture has in preserving testimonies of the past.

As it was explained in Chap. 1, places can be defined as physical spaces endowed with the capability of making inhabitants and observers perceive and share a certain historic, identitary, and symbolical texture. This texture, in reverse, is linked to the presence of physical elements. The latter serve as material "referents" connecting to symbols and to their meaning.

Each time a culture shapes its living spaces in a certain manner, it uses the material elements in its disposal to construct buildings, areas, and physical delimitations. The resulting reificated space can be considered, at least in some cases and at certain scales, as a semiotic "referent" to the systems of perceptions, decisions, and attitudes of the human group who had built it. That configuration of spaces is seen, lived in, and perceived by those human groups in obvious relationship with its surrounding context and environment. This is what makes that portion of space become the material basis of a cultural landscape.

"Historical" views, possibly no longer accessible to present-day observer, might be partially or fully perceived by adopting appropriate panoramic acquisition strategies with the use of drones. This approach might be particularly useful in order to clarify, from a perceptive, cultural, and social standpoint, the spatiotemporal phase stratification of a given landscape (Brumana et al. 2012).

As it was previously stated, the visual representation of places, like any other kind of iconographic/cartographic source in geography, provides for an (obviously selective and intentional) description of what is in the represented space. Also, it reflects the personal and cultural views of the author of the representation. The so-called "cityscape," for instance, can be thought of as an urban landscape whose view gives indications about material objects. Far beyond that, however, it may also imply and/or express narratives and meanings of social and cultural relevance (Germen 2016). This may involve very long to extremely short periods of time, up to "freezing" a single moment in which the relationship between people and places holds particular relevance through that particular (aerial) view (Figs. 3.13 and 3.14).

Fig. 3.13 The church of Nuestra Señora de los Remedios in Cholula (Puebla, Mexico) was founded by obliterating a pre-Colombian temple, on top of a Mixtecan pyramid. The shape of such building can be recognized as the hill on which the "new" religious monument is placed. Following a well-tested practice from Christian Europe, in replacing the "pagan" religion system of values, the conquistadores kept the socially perceived and solidly shared value of that particular *space* as a *sacred place*, transplanting a totally different spiritual and symbolical meaning in it. The image was taken by a DJI Phantom 3 quad-copter (Image: DCR Aerial Films, Mexico)

Fig. 3.14 Aerial image of the Neuschwanstein Castle (Bavaria, Germany) taken by a DJI Phantom 4 quadcopter. The castle was built in the late nineteenth century by King Ludwig II, as a monument to traditional Germanic sagas and myths. Although a state-of-the art modern building in its times, the castle makes wide use of material geometries and elements drawn from a Romantic reinterpretation of late-antiquity and middle-ages styles. In this sense, following Vallega (2003), Neuschwanstein can be interpreted as a specific kind of "hyper-place," or a place which points to symbols and meanings above and beyond human reality. Aerial views from small drones can be useful for instantiating, reinterpreting, and – so to speak – amplifying in an innovative manner highly symbolic and timeless landscapes (Image: Matt Edwards – LostScientist.com)

In principle, material configurations may be decoded by any sufficiently informed observer. This fact is well known to cultural and humanities-oriented geographers, who explore places and social contexts in order to draw elements of knowledge about the immaterial drivers of anthropic phenomena. Drones can indeed prove helpful tools toward that kind of understanding. On the one hand, they can contribute to making views of the artifacts more readable; on the other hand, their technical evolution as data acquisition platforms, along with GIS-based processing workflows, will allow for increasingly complex spatial and semantic analyses.

It should be underlined that humanistic and cultural geographers have identified conceptual differences in the inherent symbolic nature of spaces, depending on the type of perception that those spaces trigger in their users. "Homotopias," for example, are places in which the entire symbolic system attached to reification is substantially uniform. "Heterotopias," drawing from some concepts expressed by

Foucault (1986) and Soja (1995), are kinds of places whose inner symbolic system is different from the – assumedly – homogeneous one in the surroundings. The diversity is recognizable by the fact that physical markers within the heterotopia point to symbols that are different from those in the external areas. A heterotopia bears some kind of spatial – if not physical – boundary, and its inner system of symbols and values, linked to the referents, is in some form of dialectic with the surrounding context. At least from a purely theoretical point of view, then, it could be possible to qualitatively detect these two elements (boundary and dialectic) by acquiring close-range aerial views of the involved areas and then submit those to interpretation.

Naturally, the fact that regions within a given area show markers of a certain symbolical texture does not mean that represented symbols are always "active" and shared by the communities. If it is true that cultures progressively reificate their spaces and disseminate them with "permanent" symbolization referents, it is also true that the mere existence – and therefore view – of those permanent referents might or might not be in correspondence to the current status of the socially perceived and shared symbols. The most obvious example is when a dominant power tries to "impose" a certain symbolization by forcing the creation of referents. Beyond this, temporal evolution plays a role anyway.

Permanent referents take time to be created. A symbolical texture may be present and it might not have produced visible referents yet. Referents may also live a long physical life even after the social and cultural context which had created them has evolved. This means that if we were only to decide whether a visible shift in referents over a certain area corresponds to a *real* heterotopia, the mere evaluation of those referents would not be sufficient for an accurate interpretation. It could indeed be a heterotopia, or it could have been so in the past, but the referents have not been changed or removed for cultural heritage reasons. Another case would be if the symbolical dissimilarity persists but the tension between the two systems faded into a peaceful identity distinction.

Anyhow, in many cases, when an anthropized space is observed, referents and indicators of certain socially and culturally relevant aspects can possibly be identified. When this is the case, they concur to provide materials for more complex analyses beyond mere physical observation.

Elements other than material referents may provide information about geosocial and geocultural aspects in the life of communities. Such is the case, for instance, of public events, ceremonies, or signs, attitudes, and behaviors when visually/visibly manifested.

In this particular type of application, small UAVs retain their nature as powerful tools for data acquisition. Obviously, all of the above could be identified and studied by other pre-drone means too. Nevertheless, drones can undoubtedly prove more effective in some specific cases (Fig. 3.15).

Fig. 3.15 A "cityscape"of Budapest - taken by a DJI Phantom 4 quadcopter - echoes a story between memories of the city's royal past and its current projection toward the future as a major European metropolis (Image: Márton Déak)

References

d'Oleire-Oltmanns S, Marzolff I et al (2012) Unmanned Aerial Vehicles (UAVs) for monitoring soil erosion in Morocco. Remote Sens 4(11):3390–3416. doi:10.3390/rs4113390

Abbas S, Qamer FM et al (2010) Application of object based image analysis for forest cover assessment of moist temperate Himalayan Forest in Pakistan. Int Arch Photogramm Remote Sens Spat Inf Sci, 39–4/C7. Available via http://dfwm.ugent.be/geobia/proceedings/papers%20proceedings/abbas_18_APPLICATION%20OF%20OBJECT%20BASED%20IMAGE%20ANALYSIS%20FOR%20FOREST%20COVER%20ASSESSMENT%20OF%20MOIST%20TEMPERATE%20HIMALAYAN%20FOREST%20IN%20PAKISTAN.pdf. Accessed 21 Apr 2017

Achille C, Adami A et al (2015) UAV-based photogrammetry and integrated Technologies for Architectural Applications – methodological strategies for the after-quake survey of vertical structures in mantua (Italy). Sensors 15(7):15520–15539. doi:10.3390/s150715520

Alidoost F, Arefi H (2015) An image-based technique for 3D building Reconstruction using multi-view UAV images. Int Arch Photogramm Remote Sens Spat Inf Sci XL-1(W5):43–46. doi:10.5194/ipsrsarchives-XL-1-W5-43-2015

Alonso-Benito A, Arroyo LA et al (2016) Fusion of WorldView-2 and LiDAR data to map fuel types in the Canary Islands. Remote Sens 8:699. doi:10.3390/rs8080669

Al-Rawabdeh A, Al-Gurrani H et al (2016) A robust registration algorithm for point clouds from UAV images for change detection. Int Arch Photogramm Remote Sens Spat Inf Sci XLI-B1:765–772. doi:10.5194/isprs-archives-XLI-B1-765-2016

Amrullah C, Suwardhi D, Meilano I (2016) Product accuracy effect of oblique and vertical non-metric digital camera utilization in UAV-photogrammetry to determine fault plane. ISPRS Ann Photogramm Remote Sens Spat Inf Sci III-6:41–48. doi:10.5194/isprsannals-III-6-41-2016

Anca P, Calgaru A et al (2016) A workflow for UAV's integration into a geodesign platform. Int Arch Photogramm Remote Sens Spat Inf Sci XLI-B1:1099–1103. doi:10.5194/isprs-archives-XLI-B1-1099-2016

Arseni M, Georgescu LP, Murariu G (2016) Photogrammetric applications using UAV systems. Ann "Dunarea de Jos" Univ Galati 2(1):37–43

Ballari D, Orellana D et al (2016) UAV monitoring for environmental management in Galapagos Islands. Int Arch Photogramm Remote Sens Spat Inf Sci XLI-B1:1105–1111. doi:10.5194/isprsarchives-XLI-B1-1105-2016

Ballarin M, Buttolo V et al (2013) Integrated surveying techniques for sensitive areas: san Felice sul Panaro. ISPRS Ann Photogramm Remote Sens Spat Inf Sci II-5(W1):19–24

Balletti C, Guerra F, Vernier P (2014) Fotogrammetria da UAV per integrare il rilievo e la documentazione di monumenti colpiti da sisma. GeoMedia 1:27–29

Balletti C, Guerra F et al (2015) 3D integrated methodologies for the documentation and the virtual reconstruction of an archaeological site. Int Arch Photogramm Remote Sens Spat Inf Sci XL-5(W4):215–222. doi:10.5194/isprsarchives-XL-5-W4-215-2015

Baluja J, Diago MP et al (2012) Assessment of vineyard water status variability by thermal and multispectral imagery using an unmanned aerial vehicle (UAV). Irrig Sci 30(6):511–522. doi:10.1007/s00271-012-0382-9

Baofeng S, Jinru X et al (2016) Digital surface model applied to unmanned aerial vehicle based photogrammetry to assess potential biotic or abiotic effects on grapevine canopies. Int J Agric Biol Eng 9(6):119–130

Bareth G, Bolten A et al (2016) Comparison of uncalibrated RGBVI with spectrometer-based NDVI derived from UAV sensing systems on field scale. Int Arch Photogramm Remote Sens Spat Inf Sci XLI-B8:837–843. doi:10.5194/isprs-archives-XLI-B8-837-2016

Barreiro A, Domínguez JM et al (2014) Integration of UAV photogrammetry and SPH modelling of fluids to study runoff on real terrains. PLoS One 9(11):e111031. doi:10.1371/journal.pone.0111031

Bates TS, Quinn PK et al (2013) Measurements of atmospheric aerosol vertical distributions above Svalbard, Norway, using unmanned aerial systems (UAS). Atmos Meas Tech 6:2115–2120. doi:10.5194/amt-6-2115-2013

Bendea H, Boccardo P et al (2008) Low cost UAV for post-disaster assessment. Int Arch Photogramm Remote Sens Spat Inf Sci XXXVII(B8):1373–1379

Bendig J, Bolten A et al (2014) Estimating biomass of barley using crop surface models (CSMs) derived from UAV-based RGB imaging. Remote Sens 6(11):10395–10412. doi:10.3390/rs61110395

Bertalan L, Túri Z, Szabó G (2016) UAS photogrammetry and object-based image analysis (GEOBIA): erosion monitoring at the Kazár Badland, Hungary. Landsc Environ 10(3–4):169–178. doi:10.21120/LE/10/3-4/10

Bignante E (2010) The use of photo-elicitation in field research. EchoGéo [online] 11.10.4000/echogeo.11622. Available via http://echogeo.revues.org/11622;DOI:10.4000/echogeo.11622. Accessed 3 Mar 2017

Blaschke T, Hay GJ et al (2014) Geographic object-based image analysis – towards a new paradigm. ISPRS J Photogramm Remote Sens 87:180–191. doi:10.1016/j.isprsjprs.2013.09.014

Boccardo P, Chiabrando F et al (2015) UAV deployment exercise for mapping purposes: evaluation of emergency response applications. Sensors 15:15717–15737. 10.330/s150715717

Bollard-Breen B, Brooks JD et al (2015) Application of an unmanned aerial vehicle in spatial mapping of terrestrial biology and human disturbance in the McMurdo dry valleys, East Antarctica. Polar Biol 38(4):573–578. doi:10.1007/s00300-014-1586-7

Borra-Serrano I, Peña JM et al (2015) Spatial quality evaluation of resampled unmanned aerial vehicle-imagery for weed mapping. Sensors 15(8):19688–19708. doi:10.3390/s150819688

Bourne JK (2016) In the empty Arctic, how do you get the job done? With a drone. In: National Geographic, April 14, 2016. Available via http://news.nationalgeographic.com/2016/04/160414-Arctic-drones-wildlife-fire-oil-spill-environment/. Accessed 2 Mar 2017
Brouwer RL, de Schipper MA et al (2015) Surfzone monitoring using rotary wing unmanned aerial vehicles. J Atmos Ocean Technol 32:855–863. doi:10.1175/JTECH-D-14-00122.1
Brumana R, Oreni D et al (2012) Panoramic UAV views for landscape heritage analysis integrated with historical maps atlases. Geoinformatics FCE CTU 9:39–50. doi:10.14311/gi.9.4
von Bueren S, Burkart A et al (2014) Comparative validation of UAV based sensors for the use in vegetation monitoring. Biogeosci Discuss 11(3):3837–3864. doi:10.5194/bgd-11-3857-2014
von Bueren SK, Burkart A et al (2015) Deploying four optical UAV-based sensors over grassland: challenges and limitations. Biogeosciences 12:163–175. https://doi.org/10.5194/bg-12-163-2015
Bühler Y, Adams MS et al (2016) Mapping snow depth in alpine terrain with unmanned aerial systems (UASs): potential and limitations. Cryosphere 10:1075–1088. doi:10.5194/tc-10-1075-2016
Burgos S, Mota M et al (2015) Use of very high-resolution airborne images to analyse 3D canopy architecture of a vineyard. Int Arch Photogramm Remote Sens Spat Inf Sci XL-3(W3):399–403. doi:10.5194/isprsarchives-XL-3-W3-399-2015
Candiago S, Remondino F et al (2015) Evaluating multispectral images and vegetation indices for precision farming applications from UAV images. Remote Sens 7(4):4026–4047. doi:10.3390/rs70404026
Capolupo A, Pindozzi S et al (2014) Indirect field technology for detecting areas object of illegal spills harmful to human health: application of drones, photogrammetry and hydrological models. Geospat Health 8(3):S699–S707. doi:10.4081/gh.2014.298
Capolupo A, Kooistra L et al (2015) Estimating plant traits of grasslands from UAV-acquired hyperspectral images: a comparison of statistical approaches. Int J Geo-Inf 4(4):2792–2820. doi:10.3390/ijgi4042792
Caprioli M, Trizzino R et al (2016) Experiences of UAV surveys applied to environmental risk management. Int Arch Photogramm Remote Sens Spat Inf Sci XLI-B1:797–801. doi:10.5194/isprsarchives-XLI-B1-797-2016
Casagrande G, Salvatori F (2011) Geografia e prospezione territoriale con micro-droni. Resoconto di una sperimentazione a fini di ricerca archeologica. If Press, Morolo. (in Italian)
Casagrande G, Salvatori F et al (2015) A "Low-tech" Approach as a Strategy for effective UAV surveying of landscape and cultural Heritage. Experiences from the Lab and the Field. Documenti Geografici 2:33–65. doi:10.19246/dg.v0i2.80
Castilla G, Hay GJ (2008) Image objects and geographic objects. In: Blaschke T, Lang S, Hay GJ (eds) Object-based image analysis: spatial concepts for knowledge-driven remote sensing applications. Springer, Berlin, pp 93–112
Čermákova I, Komárková J, Sedlák P (2016) Using UAV to detect shoreline changes: case study – Pohranov Pond, Czech Republic. Int Arch Photogramm Remote Sens Spat Inf Sci XLI-B1:803–808. doi:10.5194/isprsarchives-XLI-B1-803-2016
Chen J, Liu H et al (2016) Damage degree evaluation of earthquake area using UAV aerial image. Int J Aerosp Eng 2016. doi: 10.1155/2016/2052603
Chiang KW, Tsai ML, Chu CH (2012) The development of an UAV borne direct georeferenced photogrammetric platform for ground control point free applications. Sensors 12(7):9161–9180. doi:10.3390/s120709161
Chiang KW, Tsai ML et al (2015) A new calibration method using low cost MEM IMUs to verify the performance of UAV-borne MMS payloads. Sensors 15(3):6560–6585. doi:10.3390/s150306560
Chiarini S, Cremonesi S et al (2014) A multi-range approach for cultural heritage survey: a case study in mantua Unesco site. Int Arch Photogramm Remote Sens Spat Inf Sci XL-5:157–164. doi:10.5194/isprsarchives-XL-5-157-2014
Choi K, Cheon JW et al (2016) An automatic UAV mapping system for supporting UN (United Nations) field operations. Int Arch Photogramm Remote Sens Spat Inf Sci XLI-B4:85–90. doi:10.5194/isprs-archives-XLI-B4-85-2016

Chrétien LP, Théau J, Ménard P (2015) Wildlife multispecies remote sensing using visible and thermal infrared imagery acquired from an unmanned aerial vehicle (UAV). Int Arch Photogramm Remote Sens Spat Inf Sci XL-1(W4):241–248. doi:10.5194/isprsarchives-XL-1-W4-241-2015
Christensen BR (2015) Use of UAV or remotely piloted aircraft and forward-looking infrared in forest, rural and wildland fire management: evaluation using simple economic analysis. N Z J For Sci 45(16). doi:10.1186/s40490-015-0044-9
Clark Howard B (2016) Drones spot swarms of sharks around Stunning Island. In: National Geographic, March 4th. Available via http://news.nationalgeographic.com/2016/03/160304-drone-mapping-st-joseph-atoll-seychelles-sharks-rays-turtles-drone-adventures/. Accessed 13 Apr 2017
Clements D, Dugdale T et al (2014) Detection of alligator weed using an unmanned aerial vehicle. Plant Protect Q 29(3):84–89
Constantinescu SG (2013) UAVs use for the support of emergency response teams specific missions. INCAS Bull 5(1):63–70. doi:10.13111/2066–8201.2013.5.1.7
Corrigan CE, Roberts GC et al (2007) Capturing vertical profiles of aerosols and black carbon over the Indian Ocean using autonomous unmanned aerial vehicles. Atmos Chem Phys Discuss Eur Geosci Union 7(4):11429–11463
Czapski P, Kacprzak M et al (2015) Preliminary analysis of the Forest health state based on multispectral images acquired by unmanned aerial vehicle. Folia Forestalia Pol 57(3):138–144. doi:10.1515/ffp-2015-0014
D'Alessandro A (2016) Drones: new tools for natural risk mitigation and disaster response. Curr Sci 110(6):958–959
Daakir M, Pierrot-Deseilligny M et al (2015) UAV onboard photogrammetry and GPS Positioning for earthworks. Int Arch Photogramm Remote Sens Spat Inf Sci XL-3(W3):293–303. doi:10.5194/isprsarchives-XL-3-W3-293-2015
Dandois JP, Olano M, Ellis EC (2015) Optimal altitude, overlap, and weather conditions for computer vision UAV estimates of forest structure. Remote Sens 7(10):13895–13920. doi:10.3390/rs71013895
Davis T (2016) When the Levee slide tries to hide. Point of beginning, 10 September: 10–11
De Michele C, Avanzi F et al (2016) Using a fixed-wing UAS to map snow depth distribution: an evaluation at peak accumulation. Cryosphere 10:511–522. doi:10.5194/tc-10-511-2016
Delaplace KLW, Van Coillie FMB et al (2010) Object-based assessment of tree attributes of Acacia tortilis in Bou-Hedma, Tunisia. Int Arch Photogramm, Remote Sens Spat Inf Sci XXXVIII-4/C7. Available via: https://www.researchgate.net/publication/265646840_Object-based_assessment_of_tree_attributes_of_acacia_tortilis_in_Bou-Hedma_Tunisia . Accessed 31 Aug 2017
Demarchi L, Bizzi S, Piégay H (2016) Hierarchical object-based mapping of riverscape units and in-stream mesohabitats using LiDAR and VHR imagery. Remote Sens 8:97. doi:10.3390/rs8020097
Dewez TJB, Leroux J, Morelli S (2016) Cliff collapse hazard from repeated multicopter UAV acquisitions: return on experience. Int Arch Photogramm Remote Sens Spat Inf Sci XLI-B5:805–811. doi:10.5194/isprs-archives-XLI-B5-805-2016
Di Gennaro SF, Battiston E et al (2016) Unmanned aerial vehicle (UAV)-based remote sensing to monitor grapevine leaf stripe disease within a vineyard affected by esca complex. Phytopathol Mediterr 55(2):262–275. doi:10.14601/Phytopathol_Mediterr-18312
Dronova I (2015) Object-based image analysis in wetland research: a review. Remote Sens 7:6380–6413. doi:10.3390/rs70506380
Dunford R, Michel K et al (2009) Potential and constraints of unmanned aerial vehicle technology for the characterization of Mediterranean riparian forest. Int J Remote Sens 30:138–146. doi:10.1080/01431160903023025
Eisenbeiss H, Sauerbier M (2011) Investigation of UAV systems and flight modes for photogrammetric applications. Photogramm Rec 26(136):400–421. doi:10.1111/j.1477-9730.2011.00657.x
Eling C, Klingbeil L et al (2015) UAV real-time. Data use in a lightweight direct georeferencing system. GPS World 44–59:55
Eltner A, Kaiser A et al (2016) Image-based surface reconstruction in geomorphometry – merits, limits and developments. Earth Surf Dyn 4:359–389. doi:10.5194/esurf-4-359-2016

Faiçal BS, Costa FG et al (2014) The use of unmanned aerial vehicles and wireless sensor networks for spraying pesticides. J Syst Archit 60(4):393–404. doi:10.1016/j.sysarc.2014.01.004
Fazeli H, Samadzadegan F, Dadrasjavan F (2016) Evaluating the potential of RTK-UAV for automatic point cloud generation in 3D rapid mapping. Int Arch Photogramm Remote Sens Spat Inf Sci XLI-B6:221–226. doi:10.5194/isprs-archives-XLI-B6-221-2016
Feng Q, Liu J et al (2015) UAV remote sensing for urban vegetation mapping using random forest and texture analysis. Remote Sens 7(1):1074–1094. doi:10.3390/rs70101074
Fernández Galarreta J, Kerle N, Gerke M (2015) UAV-based urban structural damage assessment using object-based image analysis and semantic reasoning. Nat Hazards Earth Syst Sci 15:1087–1101. doi:10.5194/nhess-15-1087-2015
Fernández T, Pérez JL et al (2016) Analysis of landslide evolution affecting olive groves using UAV and photogrammetric techniques. Remote Sens 8:837. doi:10.3390/rs8100837
Flener C, Vaaja M et al (2013) Seamless mapping of river channels at high resolution using mobile LiDAR and UAV-photography. Remote Sens 5(12):6382–6407. doi:10.3390/rs5126382
Foucault M (1986) Of other spaces. Diacritics 16(1):22–27
Fraser RH, Olthof I et al (2015) UAV photogrammetry for mapping and monitoring of northern permafrost landscapes. Int Arch Photogramm Remote Sens Spat Inf Sci XL-1/W4. doi:10.5194/isprsarchives-XL-1-W4-361-2015
Frith J (2017) Fixed wing drones for coastal monitoring. Available via http://www.maritime-journal.com/news101/marine-civils/marine-civils/fixed-wing-drones-for-coastal-monitoring. Accessed 3 Mar 2017
Gallay M, Eck C et al (2016) High resolution airborne laser scanning and hyperspectral imaging with a small UAV platform. Int Arch Photogramm Remote Sens Spat Inf Sci XLI-B1:823–827. doi:10.5194/isprsarchives-XLI-B1-823-2016
Germen M (2016) Alternative cityscape visualisation: drone shooting as a new dimension in urban photography. Electronic Visualisation and the Arts (EVA 2016), London. 12-14 July 2016: 150-157. Doi: 10.14236/ewic/EVA2016.31. Available via http://ewic.bcs.org/upload/pdf/ewic_eva16_ap_paper3.pdf. Accessed 20 Apr 2017
Getzin S, Nuske RS, Wiegand K (2014) Using unmanned aerial vehicles (UAV) to quantify spatial gap patterns in forests. Remote Sens 6(8):6988–7004. doi:10.3390/rs6086988
Gevaert CM, Parsello C et al (2016) Classification of informal settlements through the integration of 2D and 3D features extracted from UAV data. ISPRS Ann Photogramm Remote Sens Spat Inf Sci III-3:317–324. doi:10.5194/isprsannals-III-3-317-2016
Giordan D, Facello A et al (2014) L'uso di UAV per monitorare e gestire il territorio nel corso di eventi legati al dissesto geo-idrologico. Geomedia 18(2):24–26
Giordan D, Manconi A et al (2015) Brief communication: "The Use of Unmanned Aerial Vehicle in a Rockfall Emergency Scenario". Nat Hazards Earth Syst Sci 15:163–169. doi:10.5194/nhess-15-163-2015
Gonzalez LF, Montes GA et al (2016) Unmanned aerial vehicles (UAVs) and artificial intelligence revolutionizing wildlife monitoring and conservation. Sensors 16:97. doi:10.3390/s16010097
Grainger A (2017) Citizen observatories and the new earth observation science. Remote Sens 9(2). doi:10.3390/rs9020153
Haas F, Hilger L et al (2016) Quantification and analysis of geomorphic processes on a recultivated iron ore mine on the Italian Island of Elba using long-term ground-based LIDAR and photogrammetric SfM data by a UAV. Nat Hazards Earth Syst Sci 16:1269–1288. doi:10.5194/nhess-16-1269-2016
Hamidi M, Samadzadegan F (2015) Precise 3D geo-location of UAV images using geo-referenced data. Int Arch Photogramm Remote Sens Spat Inf Sci XL-1(W5):269–275. doi:10.5194/isprs-archives-XL-1-W5-269-2015
Harder P, Schirmer M et al (2016) Accuracy of snow depth estimation in mountain and prairie environments by an unmanned aerial vehicle. Cryosphere 10:2559–2571. doi:10.5194/tc-10-2559-2016
Hay GJ, Hemachandran B et al (2010) HEAT – Home Energy Assessment Technologies: a WEB2.0 residential waste heat analysis using GEOBIA and airborne thermal imagery. Int Arch

Photogramm, Remote Sens Spat Inf Sci XXXVIII-4/C7. Available via www.isprs.org/proceedings/XXXVIII/part1/01/01_04_Paper_178.pdf. Accessed 13 Apr 2017
Hoffman H, Nieto H et al (2016) Estimating evaporation with thermal UAV data and two-source energy balance models. Hydrol Earth Syst Sci 20:697–713. doi:10.5194/hess-20-697-2016
Holman FH, Riche AB et al (2016) High throughput field phenotyping of wheat plant height and growth rate in field plot trials using UAV based remote sensing. Remote Sens 8(12):1031. doi:10.3390/rs8121031
Honkavaara E, Saari H et al (2013) Processing and assessment of spectrometric, stereoscopic imagery collected using a lightweight UAV spectral camera for precision agriculture. Remote Sens 5(10):5006–5039. doi:10.3390/rs5105006
Honkavaara E, Hakala T et al (2016) Geometric and reflectance signature characterization of complex canopies using hyperspectral stereoscopic images from UAV and terrestrial platforms. Int Arch Photogramm Remote Sens Spat Inf Sci XLI-B7:77–82. doi:10.5194/isprsarchives-XLI-B7-77-2016
Hosseinpoor HR, Samadzadegan F, Dadrasjavan F (2016) Precise target geolocation and tracking based on UAV video imagery. Int Arch Photogramm Remote Sens Spat Inf Sci XLI-B6:243–249. doi:10.5194/isprs-archives-XLI-B6-243-2016
Huang S, Zhang Z et al (2016) Image network generation of uncalibrated UAV images with low-cost GPS data. Int Arch Photogramm Remote Sens Spat Inf Sci XLI-B3:31–37. doi:10.5194/isprs-archives-XLI-B3-31-2016
Husson E, Lindgren F, Ecke F (2014) Assessing biomass and metal contents in riparian vegetation along a pollution gradient using an unmanned aircraft system. Water Air Soil Pollut 225(6):1957. doi:10.1007/s11270-014-1957-2
Idachaba FE, Oni OO (2015) Low power VTOL UAV deployments for urban area monitoring. Proceedings of the world congress on engineering, vol 1. WCE 2015, July 1–3, 2015, London
Immerzeel WW, Kraaijenbrink PDA, Andreassen LM (2017) Use of an unmanned aerial vehicle to assess recent surface elevation change of Storbreen in Norway. Cryosphere Discuss. doi:10.5194/tc-2016-292
Inoue T, Nagai S et al (2014) Unmanned aerial survey of fallen trees in a deciduous broadleaved forest in eastern Japan. PLoS One 9(10):e109881. doi:10.1371/journal.pone.0109881
Izumida A, Uchiyama S, Sugai T (2017) Application of UAV-SfM photogrammetry and aerial LiDAR to a disastrous flood: multitemporal topographic measurement of a newly formed crevasse splay of the Kinu River, central Japan. Nat Hazards Earth Syst Sci Discuss. doi:10.5194/nhess-2017-42
Jaud M, Grasso F et al (2016) Potential of UAVs for monitoring mudflat morphodynamics (application to the seine estuary, France). ISPRS Int J Geo-Information 5(50). doi:10.3390/ijgi5040050
Johansen K, Tiede D et al (2011) Automatic geographic object based mapping of streambed and riparian zone extent from LiDAR data in a temperate rural urban environment, Australia. Remote Sens 3(6):1139–1156. doi:10.3390/rs3061139
Karpina M, Jarząbek-Rychard M et al (2016) UAV-based automatic tree growth measurement for biomass estimation. Int Arch Photogramm Remote Sens Spat Inf Sci XLI-B8:685–688. doi:10.5194/isprsarchives-XLI-B8-685-2016
Kattenborn T, Sperlich M et al (2014) Automatic single palm tree detection in plantations using UAV-based photogrammetric point clouds. Int Arch Photogramm Remote Sens Spat Inf Sci XL-3:139–144. doi:10.5194/isprsarchives-XL-3-139-2014
Khan A, Schaefer D et al (2012) Low power greenhouse gas sensors for unmanned aerial vehicles. Remote Sens 4:1355–1368. doi:10.3390/rs4051355
Kniaz (2016) Robust vision-based pose estimation algorithm for an UAV with Known Gravity Vector. Int Arch Photogramm Remote Sens Spat Inf Sci XLI-B5:63–68. doi:10.5194/isprsarchives-XLI-B5-63-2016
Kubota S, Kawai Y (2016) River maintenance management system using three-dimensional UAV data in Japan. ISPRS Ann Photogramm Remote Sens Spat Inf Sci IV-2(W1):93–98. doi:10.5194/isprs-annals-IV-2-W1-93-2016

Kurtz TH, Buckley SJ (2016) A review of hyperspectral imaging in close range applications. Int Arch Photogramm Remote Sens Spat Inf Sci XLI-B5:865–870. doi:10.5194/isprsarchives-XLI-B5-865-2016

Laliberte AS, Herrick JE et al (2010) Acquisition, orthorectification and object-based classification of unmanned aerial vehicle (UAV) imagery for rangeland monitoring. Photogramm Eng Remote Sens 6:661–672. doi:10.14358/PERS.76.6.661

Lang S, Burnett C, Blaschke T (2004) Multiscale object-based image analysis – a key to the hierarchical organization of landscapes. Ekológia (Bratislava) 23(1):148–156

Lega M, Napoli RMA (2010) Aerial infrared thermography in the surface waters contamination monitoring. Desalin Water Treat 23(1–3):141–151. doi:10.5004/dwt.2010.1988

Lega M, Kosmatka J et al (2012) Using advanced aerial platforms and infrared thermography to track environmental contamination. Environ Forensic 13:332–338. doi:10.1080/15275922.2012.729002

Lelong CCD, Burger P et al (2008) Assessment of unmanned aerial vehicles imagery for quantitative monitoring of wheat crop in small plots. Sensors 8(5):3557–3585. doi:10.3390/s8053557

Leucci G, Masini N et al (2015) Integrated archaeogeophysical approach for the study of a medieval monastic settlement in Basilicata. Open Archaeol 1:236–246. doi:10.1515/opar-2015-0014

Li Y, Gong JH et al (2012) Efficient dam break flood simulation methods for developing a preliminary evacuation plan after the Wenchuan earthquake. Nat Hazards Earth Syst Sci 12:97–106. doi:10.5194/nhess-12-97-2012

Li D, Wang X, Sun T (2016) Energy-optimal coverage path planning on topographic map for environment survey with unmanned aerial vehicles. Electron Lett 52(9):699–701. doi:10.1049/el.2015.4551

Linchant J, Lisein K et al (2015) Are unmanned aircraft systems (UASs) the future of wildlife monitoring? A review of accomplishments and challenges. Mammal Rev 45(4):239–252. doi:10.1111/mam.12046

Lo Brutto M, Garraffa A, Meli P (2014) UAV platforms for cultural heritage survey: first results. ISPRS Ann Photogramm Remote Sens Spat Inf Sci II-5:227–234. doi:10.5194/isprsannals-II-5-227-2014

Long A, Clark B (2014) A versatile eye in the landfill sky. MSW Management, September/October, p 70–73

Long N, Millescamps B et al (2016) Accuracy assessment of coastal topography derived from UAV images. Int Arch Photogramm Remote Sens Spat Inf Sci XLI-B1:1127–1134. doi:10.5194/isprs-archives-XLI-B1-1127-2016

Louiset T, Pamart A et al (2016) A shape-adjusted tridimensional reconstruction of cultural heritage artifacts using a miniature quadrotor. Remote Sens 8:858. doi:10.3390/rs8100858

Lu CH (2016) Applying UAV and photogrammetry to monitor the morphological changes along the beach in Penghu Islands. Int Arch Photogramm Remote Sens Spat Inf Sci XLI-B8:1153–1156

Lucieer A, De Jong SM, Turner D (2014) Mapping landslide displacements using structure from motion (SfM) and image correlation of multi-temporal UAV photography. Prog Phys Geogr 38(1):97–116. doi:10.1177/0309133313515293

Lukas V, Novák J et al (2016) The combination of UAV survey and Landsat imagery for monitoring of crop vigor in precision agriculture. Int Arch Photogramm Remote Sens Spat Inf Sci XLI-B8:953–957. doi:10.5194/isprs-archives-XLI-B8-953-2016

Luna I, Lobo A (2016) Mapping crop planting quality in sugarcane from UAV imagery: a pilot study in Nicaragua. Remote Sens 8:500. doi:10.3390/rs8060500

Ma Y, Wu X et al (2016) Pedestrian detection and tracking from low-resolution unmanned aerial vehicle thermal imagery. Sensors 16(4):446. doi:10.3390/s16040446

Mader D, Blaskow R et al (2015) UAV-based acquisition of 3D point cloud – a comparison of a low-cost laser scanner and SFM-tools. Int Arch Photogramm Remote Sens Spat Inf Sci XL-3(W3):335–341. doi:10.5194/isprsarchives-XL-3-W3-335-2015

Mader D, Blaskow R et al (2016) Potential of UAV-based laser scanner and multispectral camera data in building inspection. Int Arch Photogramm Remote Sens Spat Inf Sci XLI-B1:1135–1142. doi:10.5194/isprs-archives-XLI-B1-1135-2016

Maiellaro N, Zonno M, Lavalle P (2015) Laser scanner and camera-equipped UAV architectural surveys. Int Arch Photogramm Remote Sens Spat Inf Sci XL-5(W4):381–386. doi:10.5194/isprsarchives-XL-W4-381-2015

Malinverni ES, Conati Barbaro C et al (2016) UAV surveying for a complete mapping and documentation of archaeological findings. The early Neolithic site of Portonovo. Int Arch Photogramm Remote Sens Spat Inf Sci XLI-B1:1149–1155. doi:10.5194/isprs-archives-XLI-B1-1149-2016

Mandlburger G, Pfenningbauer M et al (2016) Evaluation of a novel UAV-borne topo-bathymetric laser profiler. Int Arch Photogramm Remote Sens Spat Inf Sci XLI-B1:933–939. doi:10.5194/isprsarchives-XLI-B1-933-2016

Martin PG, Moore J et al (2016) Radiological assessment on interest areas on the Sellafield nuclear site via unmanned aerial vehicle. Remote Sens 8:913. doi:10.3390/rs8110913

Máthé K, Buşoniu L (2015) Vision and control for UAVs: a survey of general methods and inexpensive platforms for infrastructure inspection. Sensors 15(7):14887–14916. doi:10.3390/s150714887

Mathews AJ, Jensen JLR (2013) Visualizing and quantifying vineyard canopy LAI using an unmanned aerial vehicle (UAV) collected high density structure from motion point cloud. Remote Sens 5(5):2164–2183. doi:10.3390/rs5052164

Mazur M (2016) Six ways drones are revolutionizing agriculture. MIT Technol Rev, July 20. Available via https://www.technologyreview.com/s/601935/six-ways-drones-are-revolutionizing-agriculture/. Accessed 23 Mar 2017

McLean JW (2015) Bathymetric mapping from a small UAV. Sea Technology, p 23–26

Mehrdad S, Satari M et al (2016) Toward real time UAVS' image mosaicking. Int Arch Photogramm Remote Sens Spat Inf Sci XLI-B1:941–946. doi:10.5194/isprs-archives-XLI-B1-941-2016

Mesas-Carrascosa F, Clavero Rumbao I et al (2014) Positional quality assessment of orthophotos obtained from sensors onboard multi-rotor UAV platforms. Sensors 14(12):22394–22407. doi:10.3390/s141333394

Mesas-Carrascosa F, Torres-Sánchez J et al (2015) Assessing optimal flight parameters for generating accurate multispectral orthomosaics by UAV to support site-specific crop management. Remote Sensors 7(10):12793–12814. doi:10.3390/rs71012793

Mesas-Carrascosa F, Notario García MD et al (2016) An analysis of the influence of flight parameters in the generation of unmanned aerial vehicle (UAV) orthomosaicks to survey archaeological areas. Sensors 16(11):1838. doi:10.3390/s16111838

Meschini A, Petrucci E et al (2014) Point cloud-based survey for cultural heritage. An experience of integrated use of range-based and image-based Technology for the san Francesco Convent in Monterubbiano. Int Arch Photogramm Remote Sens Spat Inf Sci XL-5:413–420. doi:10.5194/isprsarchives-XL-5-413-2014

Mikita T, Janata P, Surový P (2016) Forest stand inventory based on combined aerial and terrestrial close-range photogrammetry. Forests 7:165. doi:10.3390/f8080165

Minařík R, Langhammer J (2016) Use of multispectral UAV photogrammetry for detection and tracking of Forest disturbance dynamics. Int Arch Photogramm Remote Sens Spat Inf Sci XLI-B8:711–718. doi:10.5194/isprsarchives-XLI-B8-711-2016

Miřijovský J, Langhammer J (2015) Multitemporal monitoring of the Morphodynamics of a mid-mountain stream using UAS photogrammetry. Remote Sens 7:8586–8609. doi:10.3390/rs70708586

Moon H, Kim C, Lee W (2016) UAV based 3-D positioning framework for detecting locations of buried persons in collapsed disaster area. Int Arch Photogramm Remote Sens Spat Inf Sci XLI-B8:121–124. doi:10.5194/isprsarchives-XLI-B8-121-2016

Morgenthal G, Hallermann N (2014) Quality assessment of unmanned aerial vehicle (UAV) based visual inspection of structures. Adv Struct Eng 17(3):289–302. doi:10.1260/1369-4332.17.3.289

Müllerová J, Brůna J et al (2016) Does the data resolution/origin matter? Satellite, airborne and UAV imagery to tackle plant invasions. Int Arch Photogramm Remote Sens Spat Inf Sci XLI-B7:903–908. doi:10.5194/isprs-archives-XLI-B7-903-2016

Munguia R, Urzua S, Grau A (2016) Delayed monocular SLAM approach applied to unmanned aerial vehicles. PLoS One 11(12):e0167197. doi:10.1371/journal.pone.0167197

Muñoz-Nieto AL, Rodríguez-Gonzálvez P et al (2014) UAV archaeological reconstruction: the study case of Chamartin Hillfort (Avila, Spain). ISPRS Ann Photogramm Remote Sens Spat Inf Sci II-5:259–265. doi:10.5194/isprsannals-II-5-259-2014

Musić J, Orović I et al (2016) Gradient compressive sensing for image data reduction in UAV based search and rescue in the wild. Math Probl Eng 2016. doi:10.1155/2016/6827414

Na WS, Baek J (2016) Impedance-based non-destructive testing method combined with unmanned aerial vehicle for structural health monitoring of civil infrastructures. Appl Sci 7(1). doi:10.3390/app7010015

Näsi R, Honkavaara E et al (2015) Using UAV-based photogrammetry and hyperspectral imaging for mapping bark beetle damage at tree-level. Remote Sens 7(11):15467–15493. doi:10.3390/rs71115467

Nebiker S, Lack N et al (2016) Light-weight multispectral UAV sensors and their capabilities for predicting grain yield and detecting plant diseases. Int Arch Photogramm Remote Sens Spat Inf Sci XLI-B1:963–970. doi:10.5194/isprs-archives-XLI-B1-963-2016

Nedjati A, Izbirak G et al (2016) Complete coverage path planning for a multi-UAV response system in post-earthquake assessment. Robotics 5(26). doi:10.3390/robotics5040026

Neumann PP (2013) Gas source localization and gas distribution mapping with a micro-drone. Freie Universität Berlin, Fachbereich Mathematik und Informatik, Berlin

Niedzielski T, Witek M, Spallek W (2016) Observing river stages using unmanned aerial vehicles. Hydrol Earth Syst Sci 20:3193–3205. doi:10.5194/hess-20-3193-2016

Ogden LE (2013) Drone Ecol BioSci 63(9):776. doi:10.1525/bio.2013.63.9.18

Ortiz Coder P (2013) Digitalización automática del patrimonio arqueológico a partir de fotogrametría. Virtual Archaeol Rev 4(8):46–49. doi:10.4995/var.2013.4287

Ostrowski W, Hanus K (2016) Budget UAV systems for the prospection of small- and medium-scale archaeological sites. Int Arch Photogramm Remote Sens Spat Inf Sci XLI-B1:971–977. doi:10.5194/isprsarchives-XLI-B1-971-2016

Otero LD (2016) Nooks & crannies. Roads & bridges, May: 44–49

Pan Z, Lie D et al (2016) Effects of citrus tree-shape and spraying height of small unmanned aerial vehicle on droplet distribution. Int J Agric Biol Eng 9(4):45–52

Papakonstantinou A, Topouzelis K, Pavlogeorgatos G (2016) Coastline zones identification and 3D coastal mapping using UAV spatial data. Int J Geo-Inf 5(75). doi:10.3390/ijgi5060075

Parcak S, Tuttle CA (2016) Hiding in plain sight: the discovery of a new monumental structure at Petra, Jordan, using WorldView-1 and WorldView-2 satellite imagery. Am Sch Orient Res BASOR 375:35–51

Patias P, Kaimaris D et al (2013) 3D mapping of cultural heritage: special problems and best practices in extreme case-studies. ISPRS Ann Photogramm Remote Sens Spat Inf Sci II-5(W1):223–228

Pavelka K, Šedina J et al (2016) Using remote sensing and RPAS for archaeology and monitoring in western Greenland. Int Arch Photogramm Remote Sens Spat Inf Sci XLI-B1:979–983. doi:10.5194/isprsarchives-XLI-B1-979-2016

Peña JM, Torres-Sánchez J et al (2015) Quantifying efficacy and limits of unmanned aerial vehicle (UAV) technology for weed seedling detection as affected by sensor resolution. Sensors 15(3):5609–5626. doi:10.3390/s150305609

Peppa MV, Mills JP et al (2016) Accuracy assessment of a UAV-based landslide monitoring system. Int Arch Photogramm Remote Sens Spat Inf Sci XLI-B5:895–902. doi:10.5194/isprs-archives-XLI-B5-895-2016

Perks MT, Russell AJ, Large ARG (2016) Technical note: advances in flash flood monitoring using unmanned aerial vehicles (UAVs). Hydrol Earth Syst Sci 20:4005–4015. doi:10.5194/hess-20-4005-2016

Petras V, Petrasova A et al (2016) Processing UAV and LiDAR point clouds in grass GIS. Int Arch Photogramm Remote Sens Spat Inf Sci XLI-B7:945–952. doi:10.5194/isprsarchives-XLI-B7-945-2016

Pilarska M, Ostrowski W et al (2016) The potential of light laser scanners developed for unmanned aerial vehicles – the review and accuracy. Int Arch Photogramm Remote Sens Spat Inf Sci XLII-2(W2):87–95. doi:10.5194/isprs-archives-XLII-2-W2-87-2016

Popescu D, Stoican F, Ichim L (2016) Control and optimization of UAV trajectory for aerial coverage in photogrammetry applications. Adv Electr Comput Eng 16(3):99–106. doi:10.4316/AECE.2016.03014

Possoch M, Bieker S et al (2016) Multi-temporal crop surface models combined with the RGB vegetation index from UAV-based images for forage monitoring in grassland. Int Arch Photogramm Remote Sens Spat Inf Sci XLI-B1:991–998. doi:10.5194/isprs-archives-XLI-B1-991-2016

Prieto R, Méndez E et al (2014) Aplicación de UAV a la detección de fauna como soporte a los censos cinegéticos. REDIAM Rev Teledetección 41:121–131. doi:10.4995/raet.2014.2312

Qiu L (2014) Watch: can drones help save wildlife around the world?. National Geographic, November 15, 2014. Available via http://news.nationalgeographic.com/news/2014/11/141114-drones-wildlife-poaching-animals-conservation/. Accessed 24 Feb 2017

Raeva PL, Filipova SL, Filipov DG (2016) Volume computation of a stockpile – a study case comparing GPS and UAV measurements in an open pit quarry. Int Arch Photogramm Remote Sens Spat Inf Sci XLI-B1:999–1004. doi:10.5194/isprs-archives-XLI-B1-999-2016

Reiss MLL, da Rocha RS et al (2016) Data integration acquired from micro-UAV and terrestrial laser scanner for the 3D mapping of Jesuit ruins of São Miguel das Missões. Int Arch Photogramm Remote Sens Spat Inf Sci XLI-B5:315–321. doi:10.5194/isprsarchives-XLI-B5-315-2016

Renard JB, Dulac F et al (2016) LOAC: a small aerosol optical counter/sizer for ground-based and balloon measurements of the size distribution and nature of atmospheric particles – part 2: first results from balloon and unmanned aerial vehicles flights. Atmos Meas Tech 9:3673–3686. doi:10.5194/amt-9-3673-2016

Rivas Casado M, Ballesteros Gonzalez R et al (2015) Automated identification of river hydromorphological features using UAV high resolution aerial imagery. Sensors 15(11):27969–27989. doi:10.3390/s151127969

Rivas Casado M, Ballesteros Gonzalez R et al (2016) Quantifying the effect of aerial imagery resolution in automated hydromorphological river characterisation. Remote Sens 8(8):650. doi:10.3390/rs8080650

Rodríguez Canosa GR, Thomas S et al (2012) A real-time method to detect and track moving objects (DATMO) from unmanned aerial vehicles (UAVs) using a single camera. Remote Sens 4(4):1090–1111. doi:10.3390/rs4041090

Roldán J, Joossen G et al (2015) Mini-Uav based sensory system for measuring environmental variables in greenhouses. Sensors 15(2):3334–3350. doi:10.3390/s150203334

Russo M, Manferdini AM (2014) Integration of image and range-based techniques for surveying complex architectures. ISPRS Ann Photogramm Remote Sens Spat Inf Sci II-5:305–312. doi:10.5194/isprsannals-II-5-305-2014

Ryan JC, Hubbard AL et al (2015) UAV photogrammetry and structure from motion to assess calving dynamics at store glacier, a large outlet draining the Greenland ice sheet. Cryosphere 9:1–11. doi:10.5194/tc-9-1-2015

Saadatseresht M, Hashempour AH, Hasanlou M (2015) UAV photogrammetry: a practical solution for challenging mapping projects. Int Arch Photogramm Remote Sens Spat Inf Sci XL-1(W5):619–623. doi:10.5194/isprsarchives-XL-1-W5-619-2015

Sakr M, Lari Z, El-Sheimy N (2016) Design and implementation of a low-cost UAV-based multisensor payload for rapid-response mapping applications. Int Arch Photogramm Remote Sens Spat Inf Sci XLI-B1:1017–1022. doi:10.5194/isprsarchives-XLI-B1-1017-2016

Salamí E, Barrado C, Pastor E (2014) UAV flight experiments applied to the remote sensing of vegetated areas. Remote Sens 6(11):11051–11081. doi:10.3390/rs61111051

Salonia A, Leti Messina T et al (2011) Photo scanner 3D survey for monitoring historical monuments. The case history of Porta Praetoria in Aosta. Geoinformatics FCE CTU 6:314–322. doi:10.14311/gi.6.39

Sánchez-García J, García-Campos JM et al (2016) An intelligent strategy for tactical movements of UAVs in disaster scenarios. Int J Distrib Sensors Netw 2016:8132812. doi:10.1155/2016/8132812

Sari NM, Kushardono D (2015) Object segmentation on UAV photo data to support the provision of rural area spatial information. Forum Geogr 29(1):49–58. doi:10.23917/forgeo.v29i1.792

Saur G, Krüger W (2016) Change detection in UAV video mosaics combining a feature based approach and extended image differencing. Int Arch Photogramm Remote Sens Spat Inf Sci XLI-B7:557–562. doi:10.5194/isprs-archives-XLI-B7-557-2016
Scarelli FM, Cantelli L et al (2016) Natural and anthropogenic coastal system comparison using DSM from a low cost UAV survey (Capão Novo, RS/Brazil). J Coast Res 75:1232–1236. doi:10.2112/SI75-247.1
Schirrmann M, Giebel A et al (2016) Monitoring agronomic parameters of winter wheat crops with low-cost UAV imagery. Remote Sens 8:706. doi:10.3390/rs8090706
Shi Y, Thomasson JA et al (2016) Unmanned aerial vehicles for high-throughput phenotyping and agronomic research. PLoS One 11(7):e0159781. doi:10.1371/journal.pone.0159781
Smigaj M, Gaulton R et al (2015) UAV-borne thermal imaging for forest health monitoring: detection of disease-induced canopy temperature increase. Int Arch Photogramm Remote Sens Spat Inf Sci XL-3(W3):349–354. doi:10.5194/isprsarchives-XL-3-W3-349-2015
Smith NG, Passone L et al (2014) Drones in archaeology: integrated data capture, processing, and dissemination in the al-Ula Valley, Saudi Arabia. Near East Archaeol 77(3):176–181
Soja EW (1995) Heterotopologies: a remembrance of other spaces, the citadel-LA. In: Watson S, Gibson K (eds) Postmodern cities and spaces. Blackwell, Oxford, pp 14–34
Sona G, Passoni D et al (2016) UAV multispectral survey to map soil and crop for precision farming applications. Int Arch Photogramm Remote Sens Spat Inf Sci XLI-B1:1023–1029. doi:10.5194/isprs-archives-XLI-B1-1023-2016
Stafford JV (2000) Implementing precision agriculture in the 21st century. J Agric Eng Res 76(3):267–275. doi:10.1006/jaer.2000.0577
Sterligov B, Cherkasov S (2016) Reducing magnetic noise of an unmanned aerial vehicle for high quality magnetic surveys. Int J Geophys 2016:4098275. doi:10.1155/2016/4098275
Straková H, Šedina J, Bílá Z (2015) Monitoring of heaps using various technologies. Civ Eng J 2. doi:10.14311/CEJ.2015.02.0011
Sun Z, Cao YK (2015) Data processing workflows from low-cost digital survey to various applications: three case studies of Chinese historic architecture. Int Arch Photogramm Remote Sens Spat Inf Sci XL-5(W7):409–416. doi:10.5194/isprsarchives-XL-5-W7-409-2015
Sun J, Jian Y, Wen C (2016) A camera-based target detection and positioning UAV system for search and rescue (SAR) purposes. Sensors 16(11):1778. doi:10.3390/s16111778
Suziedelyte Visockiene J, Puziene R et al (2016) Unmanned aerial vehicles for photogrammetry: analysis of ortophoto images over the territory of Lithuania. Int J Aerosp Eng Hindawi Publ Corp 2016:4141037. doi:10.1155/2016/4141037
Tahar KN (2015) Multi rotor UAV at different altitudes for slope mapping studies. Int Arch Photogramm Remote Sens Spat Inf Sci XL-1(W4):9–16. doi:10.5194/isprsarchives-XL-1-W4-9-2015
Tampubolon W, Reinhardt W (2016) Direct georeferencing of UAV data based in simple building structures. Int Arch Photogramm Remote Sens Spat Inf Sci XLI-B1:1043–1050. doi:10.5194/isprsarchives-XLI-B1-1043-2016
Tanzi TJ, Chandra M et al (2016) Towards "drone-borne" disaster management: future application scenarios. ISPRS Ann Photogramm Remote Sens Spat Inf Sci III-8:181–189. doi:10.5194/isprsannals-III-8-181-2016
Tilly N, Kelterbaum D, Zeese R (2016) Geomorphological mapping with terrestrial laser scanning and UAV-based imaging. Int Arch Photogramm Remote Sens Spat Inf Sci XLI-B5:591–597. doi:10.5194/isprs-archives-XLI-B5-591-2016
Tokarczyk P, Leitao JP et al (2015) High-quality observation of surface imperviousness for urban runoff modelling using UAV imagery. Hydrol Earth Syst Sci 19:4215–4228. doi:10.5194/hess-19-4215-2015
Tong X, Liu X et al (2015) Integration of UAV-based photogrammetry and terrestrial laser scanning for the three-dimensional mapping and monitoring of open-pit mine areas. Remote Sens 7(6):6635–6662. doi:10.3390/rs70606635
Torres-Sánchez J, López-Granados F et al (2013) Configuration and specifications of an unmanned aerial vehicle (UAV) for early site specific weed management. PLoS One 8(3):e358210. doi:10.1371/journal.pone.0058210

Tscharf A, Rumpler M et al (2015) On the use of UAVs in mining and archaeology – geo-accurate 3D reconstructions using various platforms and terrestrial views. ISPRS Ann Photogramm Remote Sens Geospat Inf Sci II-1(W1):15–22. doi:10.5194/isprsannals-II-1-W1-15-2015
Turner D, Lucieer A et al (2014) Spatial coregistration of ultra-high resolution visible, multispectral and thermal images acquired with a micro-UAV over Antarctic moss beds. Remote Sens (Basel) 6(5):4003–4024. https://doi.org/10.3390/rs6054003
Turner D, Lucieer A, de Jong SM (2015) Time series analysis of landslide dynamics using an unmanned aerial vehicle (UAV). Remote Sens 7(2):1736–1757. doi:10.3390/rs70201736
Unger J, Reich M, Heipke C (2014) UAV-based photogrammetry: monitoring of a building zone. Int Arch Photogramm Remote Sens Spat Inf Sci XL-5:601–606. doi:10.5194/isprsarchives-XL-5-601-2014
Uribe P, Angás J et al (2015) Aerial mapping and multi-sensors approaches from remote sensing applied to the roman archaeological heritage. Int Arch Photogramm Remote Sens Spat Inf Sci XL-5(W4):461–467. doi:10.5194/isprsarchives-XL-5-W4-461-2015
Vallega A (2003) Geografia culturale. Luoghi, spazi, simboli. UTET, Torino
Van Iersel WK, Straatsma MW et al (2016) Monitoring phenology of floodplain grassland and herbaceous vegetation with UAV imagery. Int Arch Photogramm Remote Sens Spat Inf Sci XLI-B7:569–571. doi:10.5194/isprsarchives-XLI-B7-569-2016
Vander Jagt B, Lucieer A et al (2015) Snow depth retrieval with UAS using photogrammetric techniques. Geosciences 5:264–285. doi:10.3390/geosciences5030264
Villa TF, Gonzalez F et al (2016a) An overview of small unmanned aerial vehicles for air quality measurements: present applications and future Prospectives. Sensors 16(7):1072. doi:10.3390/s16071072
Villa TF, Salimi F et al (2016b) Development and validation of a UAV based system for air pollution measurements. Sensors 16(12):2202. doi:10.3390/s16122202
Vrublová KR et al (2015) Documentation of landslides and Inaccessible parts of a mine using an unmanned UAV system and methods of digital terrestrial photogrammetry. GeoSci Eng 61(3):8–19. doi:10.1515/gse-2015-0018
Wallace L, Lucieer A et al (2016) Assessment of Forest structure using two UAV techniques: a comparison of airborne laser scanning and structure from motion (SfM) point clouds. Forests 7(3):62. doi:10.3390/f7030062
Wang R, Zhang S et al (2016) Gully erosion mapping and monitoring at multiple scales based on multi-source remote sensing data of the Sancha River catchment, Northeast China. Int J Geo-Inf 5(11):200. doi:10.3390/ijgi5110200
Whitehead K, Moorman, Hugenholtz CH (2013) Low-cost, on-demand aerial photogrammetry for glaciological measurement. Cryosphere Discuss 7:3043–3057. doi:10.5194/tcd-7-3043-2013
Wierzbicki D, Kedzierski M, Fryskowska A (2015) Assessment of the influence of UAV image quality on the Ortophoto production. Int Arch Photogramm Remote Sens Spat Inf Sci XL-1(W4):1–8. doi:10.5194/isprsarchives-XL-1-W4-1-2015
Wilner FN (2015) Look! Above the train! It's a bird! It's a plane! No, it's a drone. Railway Age, p 10–11
Wing MG, Burnett J et al (2013) Eyes in the sky: remote sensing technology development using small unmanned aircraft systems. J For 111(5):341–347. doi:10.5849/jof.12-117
Xu Z, Wu L et al (2014) Tridimensional reconstruction applied to cultural heritage with the use of camera-equipped UAV and terrestrial laser scanner. Remote Sens 6(11):10413–10434. doi:10.3390/rs61110413
Xu J, Solmaz G et al (2016) Internet of things applications: animal monitoring with unmanned aerial vehicle. arXiv:1610.05287v2. Available via https://arxiv.org/pdf/1610.05287.pdf. Accessed 2 Sept 2017
Xu Z, Wu TH et al (2016b) Three dimensional reconstruction of large cultural heritage objects based on UAV video and TLS data. Int Arch Photogramm Remote Sens Spat Inf Sci XLI-B5:985–988. doi:10.5194/isprsarchives-XLI-B5-985-2016
Xu Y, Yu G et al (2016c) A hybrid vehicle detection method based on Viola-Jones and HOG + SVM from UAV images. Sensors 16(8):1325. doi:10.3390/s16081325

Yanagi H, Chikatsu H (2015) Camera calibration in 3D modelling for UAV application. Int Arch Photogramm Remote Sens Spat Inf Sci XL-4(W5):223–226. doi:10.5194/isprsarchives-XL-4-W5-223-2015

Yanagi H, Chikatsu H (2016) Performance evaluation of 3D modeling software for UAV photogrammetry. Int Arch Photogramm Remote Sens Spat Inf Sci XLI-B5:147–152. doi:10.5194/isprsarchives-XLI-B5-147-2016

Ye D, Liao M et al (2016) Research on reef bathymetric survey of UAV stereopair based on two-medium photogrammetry. Int Arch Photogramm Remote Sens Spat Inf Sci XLI-B1:407–412. doi:10.5194/isprs-archives-XLI-B1-407-2016

Yeh ML, Chou YT, Yang LS (2016) The evaluation of GPS techniques for UAV-based photogrammetry in urban area. Int Arch Photogramm Remote Sens Spat Inf Sci XLI-B1:1079–1084. doi:10.5194/isprsarchives-XLI-B1-1079-2016

Yoo CI, Oh TS (2016) Beach volume change using UAV photogrammetry Songjung Beach, Korea. Int Arch Photogramm Remote Sens Spat Inf Sci XLI-B8:1201–1205. doi:10.5194/isprs-archives-XLI-B8-1201-2016

Yu H, Yan T et al (2016) An integrative object-based image analysis workflow for UAV images. Int Arch Photogramm Remote Sens Spat Inf Sci XLI-B1:1085–1091. doi:10.5194/isprsarchives-XLI-B1-1085-2016

Yuan Y, Hu X (2016) Random forest and objected-based classification for forest pest extraction from UAV aerial imagery. Int Arch Photogramm Remote Sens Spat Inf Sci XLI-B1:1093–1098. doi:10.5194/isprs-archives-XLI-B1-1093-2016

Yuan Y, Wang Z et al (2015) DDDAS-based information-aggregation for crowd dynamics modeling with UAVs and UGVs. Front Robot AI 2:1–10. doi:10.3389/frobt.2015.00008

Zarco-Tejada P, González-Dugo V, Berni J (2011) Fluorescence, temperature and narrow-band indices acquired from a UAV platform for water stress detection using a micro-hyperspectral imager and a thermal camera. Remote Sens Environ 117:322–337. doi:10.1016/j.rse.2011.10.007

Zarnowski A, Banaszek A, Banaszek S (2015) Application of technical measures and software in constructing photorealistic 3D models of historical building using ground-based and aerial (UAV) digital images. Rep Geodesy Geoinformatics 99:54–63. doi:10.2478/rgg-2015-0012

Zawieska D, Markiewicz J et al (2016) Multi-criteria GIS analyses with the use of UAVs for the needs of spatial planning. Int Arch Photogramm Remote Sens Spat Inf Sci XLI-B1:1165–1171. doi:10.5194/isprs-archives-XLI-B1-1165-2016

Zhang C, Kovacs JM (2012) The application of small unmanned aerial systems for precision agriculture: a review. Precis Agric 13(6):693–712. https://doi.org/10.1007/s11119-012-9274-5

Zmarz A (2014) UAV – a useful tool for monitoring woodlands. Misc Geograph Reg Stud Dev 18(2):46–52. doi:10.2478/mgrsd-2014-0006

Zmarz A, Korczak-Abshire M et al (2015) Indicator species population monitoring in Antarctica with UAV. Int Arch Photogramm Remote Sens Spat Inf Sci XL-1/W4:189–193. doi:10.5194/isprsarchives-XL-1-W4-189-2015

Chapter 4
Zooming on Aerial Survey

Gergely Szabó, László Bertalan, Norbert Barkóczi, Zoltán Kovács, Péter Burai, and Csaba Lénárt

Abstract The aim of this chapter is to provide a general overview about the main components of a developed UAS mapping system, the survey, and processing procedure. At first (4.1), a brief introduction is given about basic operational elements and accessories of UAS. Then, recent camera/sensor technologies allowing various survey solutions are going to be discussed. Once these hardware components are presented, the detailed workflow of a basic UAV-based mapping procedure is described (4.2). A further discussion focuses not only on the analytical or planning phases but also on providing useful information on the operational and processing parts as well (4.3). Then, there comes image acquisition and project planning (4.4). The photogrammetry-based image processing requires detailed expertise and attention; Sect. 4.5 maybe helpful to avoid potential mistakes. The last section (4.6) summarizes some aspects of the use of LiDAR technologies in UAV-based surveys.

4.1 Different Instrument Types

Gergely Szabó – László Bertalan – Norbert Barkóczi

Once the analytical and operational phases of the UAV mapping are clarified, then the image acquisition part follows. However, various types of sensor payloads are available for different research purposes. Each one has crucial application requirements to be fulfilled. It is usually demanding to clarify the optimal combination of the UAV and the possible type of devices to be carried. Basically, in the case of UAS (especially for low-cost versions), the vehicle configuration is predefined by the manufacturer. Consequently, fitting the diversity of existing remote sensing payloads on off-the-shelf products is often challenging for the operator or for the end user (Colomina and Molina 2014). The vehicles generally have specific volume, weight, and power specifications. Thus, if the aim is to serve the specific application requirements (remote sensing bandwidth, accuracy, spatial/spectral resolution, etc.),

---

G. Szabó (✉) • L. Bertalan • N. Barkóczi • Z. Kovács • P. Burai • C. Lénárt
Department of Physical Geography and GIS, University of Debrecen, Debrecen, Hungary
e-mail: szabo.gergely@science.unideb.hu; norbert.barkoczi@gmail.com; bertalanlaszlo.geo@gmail.com; kovacszoli88.geo@gmail.com; pburai@gmail.com; lenart.dr@gmail.com

G. Casagrande et al. (eds.), *Small Flying Drones: Applications for Geographic Observation*, DOI 10.1007/978-3-319-66577-1_4

then it is frequently problematic to find a right balance. Moreover, it is also troublesome that the existing camera/sensor systems cover a large area from low-cost to professional categories. Plus, finally, most of them are specifically designed for only one type of UAV system.

4.1.1 RGB Visible-Band, RGB Near-Infrared, and Multispectral Cameras

Recently, the mass market produced each year higher and higher resolution sensors/cameras with smaller pixel size. This has obviously facilitated developments in our specific sector of remote sensing as well. The majority of the low-cost UAV mapping investigations focus on the ability to produce high-resolution digital orthophotographs and digital surface models. These are supposed to be base data for further analysis. Visible-band sensors are widely used for this purpose; however, during the analytical phase, the following parameters need detailed consideration: sensor type and resolution, pixel size, frame rate, shutter speed, focal length, and even the total weight of the camera/lens system. All of this should be clarified in the beginning (Georgopoulos et al. 2016). Generally, UAV mapping workflows are based on sensors such as the CCD and CMOS DSLR (digital single-lens reflex) cameras. However, mirrorless interchangeable-lens cameras (MILC) became increasingly popular in the past few years since they offer light-weight solutions. Fixed-wing UASs (unmanned aerial systems – not only the aircraft, but the controller and all the necessary software as well) generally carry built-in optical cameras. These often provide poor resolution and modest precision. Thus, while these UAVs are capable of long flight durations, in some cases they produce unacceptable results. Table 4.1 summarizes some of the most common optical cameras used for UAV mapping.

In contrast to the basic aerial imaging systems, novel methodologies arose to decrease the flight time and increase the achievable angles at which images were to be acquired. A possible solution is to combine several cameras in one UAV. Xie et al. (2012) designed a wide-angle camera system that is built up from four Canon EOS 5D Mark II cameras to reach a wide (130°) angle in total. A self-calibration procedure, invented for this system, manages the overlaps of the four single images in order to eliminate errors caused by mechanical deformations and by the time difference in exposure. A five-camera system was tested by Grenzdörffer et al. (2012), but these authors found that the UAV motion caused inaccuracies in image acquisition.

Range imaging sensors (RIM) hold a special place in the set of optical sensors. Cameras of this type of camera use the time-of-flight (ToF) principle and an amplitude-modulated continuous light emitter along with a CCD/CMOS receiver (Kohoutek and Eisenbeiss 2012). The reflected light is sampled four times regularly by the sensor with an internal phase delay. The correlation amplitude is then calculated along with the incident light intensity. 3D coordinates are extracted by deriving the measured distance from these values; however, raw images are to be thoroughly processed.

In some cases, a cost-reducing solution could be the use of high-level smartphones having great quality built-in camera sensors to produce digital elevation

Table 4.1 Common and/or representative optical cameras suitable for UAS

Manufacturer and model	Sensor type resolution (MPx)	Format type	Sensor size (mm^2)	Pixel pitch (μm)	Weight (kg)	Frame rate (fps)	Max shutter speed (s^{-1})	Approx. price ($)
Canon EOS 5DS	CMOS 51	FF	36.0×24.0	4.1	0.930	5.0	8000	3400
Sony Alpha 7R II	CMOS 42	FF MILC	35.9×24.0	4.5	0.625	5.0	8000	3200
Pentax 645D	CCD 40	FF	44.0×33.0	6.1	1.480	1.1	4000	3400
Nikon D750	CMOS 24	FF	35.9×24.0	6.0	0.750	6.5	4000	2000
Nikon D7200	CMOS 24	SF	23.5×15.6	3.9	0.675	6.0	8000	1100
Sony Alpha a6300	CMOS 24	SF MILC	23.5×15.6	3.9	0.404	11.0	4000	1000
Pentax K-3 II	CMOS 24	SF	23.5×15.6	3.9	0.800	8.3	8000	800
Canon EOS 7D Mark II	CMOS 20	SF	22.3×14.9	4.1	0.910	10.0	8000	1500
Panasonic Lumix DMC GX8	CMOS 20	SF MILC	17.3×13.0	3.3	0.487	10.0	8000	1000
Ricoh GXR A16	CMOS 16	SF	23.6×15.7	4.8	0.550	2.5	3200	650

Adapted from Georgopoulos et al. (2016)

models (e.g., Samsung Galaxy S2, Yun et al. 2012; Nokia Lumia 1020, Lehmann Aviation LA300 UAS).

Visible-range optical cameras and sensors have several limitations for detailed Earth observations since many variables are just not visible in a basic aerial image. If it is possible to reach a higher section of the electromagnetic spectrum, then more of them become observable. A good example can be found in studies focusing on vegetation analysis, where the near-infrared (NIR) range is needed to reveal changes of health variables by the calculated VI (Vegetation Index) values.

The multiple-camera array by Tetracam was one of the first sensors applicable on UAV systems. The multispectral sensor offered three different versions. Each contains 4, 6, or 12 factory-aligned cameras in a multispectral set. Each camera contains a customer-specified narrow-band filter that is inserted between the lens and sensor. With each exposure, 4, 6, or 12 separate bands of visible or near-infrared radiation pass through each camera's lens and filter to form a separate monochromatic image on the camera's sensor (www.tetracam.com).

The newer multi-head sensors (e.g., Parrot Sequoia; Fig. 4.1) combine high-resolution RGB cameras and individual multispectral sensors in order to increase the strength of both systems. As a result, they provide for high spatial resolution, eligible for bundle adjustment and extraction of geometric parameters. These cameras offer individual multispectral sensors equipped with high-class interference

Fig. 4.1 The most popular multispectral sensors for UAVs: (**a**) Tetracam MiniMCA-6; (**b**) Parrot Sequoia; (**c**) MicaSense RedEdge

Table 4.2 Common and/or representative multispectral cameras for UAS

Manufacturer and model	Resolution (Mpx)	Size (mm)	Pixel size (μm)	Weight (kg)	Number of spectral bands	Spectral range (nm)
Tetracam MiniMCA-6	1.3	131×78×88	5.2×5.2	0.7	6	450–1000
Tetracam ADC micro	3.2	75×59×33	3.2×3.2	0.9	6	520–920
Quest Innovations Condor-5 ICX 285	7	150×130×177	6.45×6.45	1.4	5	400–1000
Parrot Sequoia	1.2	59×41×28	3.75×3.75	0.72	4	550–810
MicaSense RedEdge		120×66×46		0.18	5	475–840

Modified after Colomina and Molina (2014)

filters. These ensure high-precision spectral measurements that could replace the ground-based spectral reference measurements in the future.

The advantages of UAV-installed multispectral cameras are evident in precision farming, as UAVs are able to produce low-cost and reliable Vegetation Index maps, showing the healthy and unhealthy areas of a cultivated field (Candiago et al. 2015). These solutions guarantee flexibility in obtaining fast and low-cost generation of VIs, as long as it is possible to associate data from the survey with calibration information and data about meteorological variables. In general, though, these high-resolution aerial images may replace ground inspections of farmlands in the near future (Nebiker et al. 2016). Table 4.2 summarizes some of the most common multispectral cameras used for UAV mapping.

4.1.2 *Hyperspectral Cameras*

Moving forward from the lower spectral resolutions and the discrete bands provided by the multispectral cameras, the hyperspectral sensors allow to acquire remote sensing imagery at narrow spectral bands over a continuous range. Consequently,

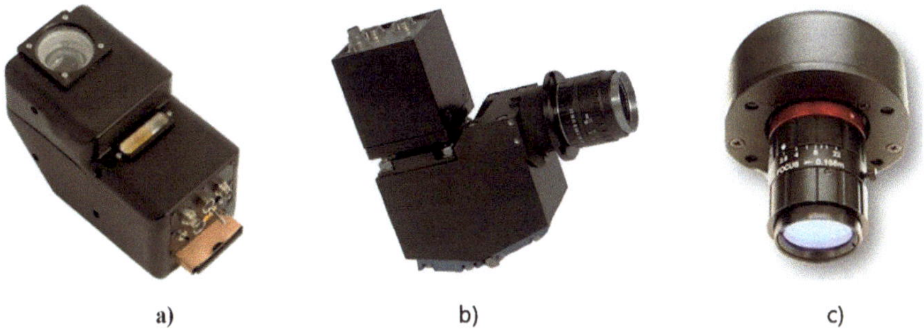

Fig. 4.2 The most popular hyperspectral sensors for UAVs: (**a**) Rikola hyperspectral camera; (**b**) Hyperspec X-series NIR camera; (**c**) OCI-UAV-1000 camera

Table 4.3 Common and/or representative hyperspectral cameras for UAS

Manufacturer and model	Lens	Size (mm^2)	Pixel size (μm)	Weight (kg)	Spectral range (nm)	Spectral bands and resolution
Rikola Ltd. hyperspectral camera	CMOS	5.6×5.6	5.5	0.6	500–900	40 10 nm
Headwall Photonics Micro-hyperspec X-series NIR	InGaAs	9.6×9.6	30	1.025	900–1,700	62 12.9 nm
BaySpec's OCI-UAV-1000	C-mount	~10×10	N/A	0.63	600–1,000	100 5 nm

Modified after Colomina and Molina (2014)

hyperspectral sensors (see examples on Fig. 4.2) are capable of extracting more details, in comparison to multispectral cameras. The reason is that each one of the recorded pixels covers the entire spectrum. Multispectral and hyperspectral sensors, however, bear an important difference from cameras operating in the visible. Optical sensors/cameras for visible spectrum analysis developed well in the last few years: systems reached tens of megapixels within containers of a few hundred grams weight. Miniaturization of multi- and hyperspectral cameras implies a whole different story, especially as far as sensor calibration processes are concerned (Colomina and Molina 2014).

Moreover, flight motions and vibrations affect the push-broom sensor mounted on the UAV, degrading the raw hyperspectral data causing distortions. Consequently, robust geometric corrections have to be performed at each single scan line, and these applications are still under development for a wider applicability, particularly in the agricultural sector (Kalisperakis et al. 2015).

Table 4.3 summarizes some of the most common hyperspectral cameras used for UAV mapping.

Fig. 4.3 Examples from thermal imaging cameras: (**a**) FLIR Vue 640; (**b**) Workswell WIRIS; (**c**) Thermoteknix Miricle

4.1.3 Thermal Cameras

Recently, miniaturization of thermal infrared sensors led to their wide applicability on unmanned platforms (Colomina and Molina 2014). As it was previously stated, UAV systems are flexible: their integration with thermal imaging solutions brings about possibilities for real-time surveillance, forest fire monitoring (Scholtz et al. 2011), risk management, detection of survivors in rescue missions (Levin et al. 2016), calculation of energy efficiency, or thermal loss of buildings (Carrio et al. 2015). The acquired thermal imagery allows for per-pixel high-precision temperature measurements. From these, 3D "thermal" models can be also reconstructed. An important application can be found in cultural heritage surveys, when previously undocumented architectural features can be revealed by noninvasive thermal acquisition (Georgopoulos et al. 2016).

The novelty of present infrared technologies relies on the concept of uncooled microbolometers as thermal sensors which are less sensitive than the cooled thermal and photon detector imagers with higher noise factor (Fig. 4.3). Current thermal imagery offers much lower resolution than optical RGB images: typically, on small UAVs, current achievable image size is 320×256, or 384×288, or 640×512 pixels at best.

Thermal imagery requires special georeferencing procedures, unlike optical imagery solutions. Basically, it is difficult to find natural GCPs in lower-resolution thermal images; thus, artificial control points are needed, mainly using aluminum as material. Aluminum GCPs show a sharp boundary in the thermal image, and automatized identification algorithms already exist for them.

Table 4.4 summarizes some of the most common thermal cameras used for UAV mapping.

4.1.4 Laser Scanners

Nowadays, terrestrial laser scanning (TLS) and airborne laser scanning (ALS) – sometimes collectively referred to as Light Detection and Ranging (LiDAR) – are common and widely used methods in photogrammetry accompanied by medium or large-format optical cameras. Still, their utilization for UAV systems is still challenging (Colomina and Molina 2014) in terms of size and weight (Fig. 4.4).

Table 4.4 Common and/or representative thermal cameras suitable for UAS

Manufacturer and model	Resolution (Px)	Sensor size (mm^2)	Pixel pitch (μm)	Weight (kg)	Spectral range(μm)	Thermal sensitivity (mK)
FLIR Vue Pro 640	640×512	10.8×8.7	17	<0.115	7.5–13.5	50
FLIR Vue Pro 336	336×256	5.7×4.4	17	<0.115	7.5–13.5	50
FLIR Tau2 640	640×512	N/A	17	<0.112	7.5–13.5	50
FLIR Tau2 336	336×256	N/A	17	<0.112	7.5–13.5	50
Thermoteknix Miricle 307 K	640×480	16.0×12.0	25	<0.170	8.0–12.0	50
Thermoteknix Miricle 110 K	384×288	9.6×7.2	25	<0.170	8.0–12.0	50/70
Workswell WIRIS 640	640×512	16.×12.8	25	<0.400	7.5–13.5	30/50
Workswell WIRIS 336	336×256	8.4×6.4	25	<0.400	7.5–13.5	30/50

Adapted from Georgopoulos et al. (2016)

Fig. 4.4 Examples of UAV-borne laser scanners: (**a**) Riegl VQ-820-GU; (**b**) Velodyne HDL-32E; (**c**) Velodyne VLP-16

Traditional LiDAR sensors are active remote sensing devices. They are mounted on manned aircraft and are capable of creating 3D point clouds by measuring time delays in laser pulse returns directed at target. The increasing popularity of low-cost UAVs resulted in the development of miniaturized laser scanner devices as well. The flash LiDAR system has an analogue connection with a camera and a flashbulb. Its operation is based on time delays between the "departure" of a flash illumination and its reflection from the target (Zhou et al. 2012). UAV-mounted laser scanners require the installation of several supplementary accessories, i.e., inertial measurement unit or IMU, GPS antenna, computer, sensor battery, sensor frame, etc.

Thus, fairly large and solid platform must be designed for this purpose.

Results obtained by Wallace et al. (2012) related to tree canopy studies show that increased flying altitude decreases return intensity from the top of the canopy; thus, it is not recommended to fly above 50 m (~164 ft). High scan angles should also be avoided. Furthermore, based on the spatial accuracy of the output point clouds, it seems that combined forward-backward transect paths may restrict the scan angles within lower altitude flights.

Table 4.5 summarizes some of the most common laser scanners used for UAV mapping.

Table 4.5 Common and/or representative laser scanners for UAS

Manufacturer and model	Scanning pattern	Range (m)	Weight (kg)	Angular res. (deg)	FOV (deg)	Laser class and λ (nm)	Frequency (kp/s)	Application
ibeo Automotive Systems IBEO LUX	4 Scanning parallel lines	200	1	(H) 0.125 (V) 0.8	(H) 110 (V) 3.2	Class A 905	22	A
Velodyne HDL-32E	32 Laser/detector pairs	100	2	(H) – (V) 1,33	(H) 360 (V) 41	Class A 905	700	MM
RIEGL VQ-820-GU	1 Scanning line	≥1000	25.5	(H) 0.01 (V) N/A	(H) 60 (V) N/A	Class 3B 532	200	H
Hokuyo UTM-30LX-EW	1,080 distances in a plane	30	0.37	(H) 0.25 (V) N/A	(H) 270 (V) N/A	Class 1905	200	A

Modified after Colomina and Molina (2014)

4.1.5 Synthetic Aperture Radar

The radar-based remote sensing technology is a traditional method implemented by satellite systems. In spite of not being fully implemented in UAV version, work is in progress toward this result. The main problem of the concept is that this kind of survey is mainly influenced by the diverse weather conditions.

One of the first developments by Rosen et al. (2007) was called UAVSAR, and it was designed to carry L-band antenna, but the majority of the recent UAV-based SAR systems are operating with antennas of X- or Ka-bands. The SARVANT project by Remy et al. (2012) was established at first for sensing at P-(repeat-pass) and X-(single-pass) bands. Recently, the combined P- and X-band radar interferometry provided the most accurate DSMs (digital surface model) and DEMs (digital elevation models), especially over forested areas. This system was able to conduct a survey with a rate of 100 km^2/hour, with a 0.5 m resolution at X-band and a 1.5 m at P-band. The main disadvantage was that the described system could not be installed on small drones, as its weight was 30 kg. Additionally, the required operational flight altitude was set to about 2,000 m with a ground swath of 3.5 km.

Another concept is to have millimeter-wave radars combined with the MIIMIC (microwave/millimeter-wave monolithic-integrated circuits) technology. The issues regarding antenna miniaturization seem to be solved (Essen et al. 2012). The millimeter-wave radar is sensitive for small-scale scattering surface changes that are not visible for simple radars like X-band. A wide range of applications can be investigated by this technology. Moreover, its low-noise performance makes the transmit power reducible without signal loss to noise ratio so the expensive transmitter accessories can be omitted, all good reasons to look at miniaturizing these devices for installation on standard small UAVs.

SAR sensors provide most impressive results; however, their very interpretation is still complicated and requires a different kind of expertise than optical or multispectral imagery.

Table 4.6 summarizes some of the most common synthetic aperture radars used for UAV mapping.

Table 4.6 Common and/or representative synthetic aperture radars for UAS

Manufacturer and mode	Spectral bands	Weight (kg)	Transmitted power (W)	Resolution (m)
IMSAR NanoSAR B	X and Ku	1.58	1	Between 0.3 and 5
Fraunhofer FHR MIRANDA	W	N/A	0.1	0.15
NASA JPL UAVSAR	L	200	2000	2
SELEX Galileo PicoSAR	X	10	–	1
OrbitSAR UAV-SAR	X and P	30	30 (X), 50 (P)	0.5 (X), 1.5 (P)
SARVANT UAS-System SAR	X and P	30	30 (X), 50 (P)	0.5 (X), 1.5 (P)

Modified after Colomina and Molina (2014)

4.2 Mapping with UAVs

Gergely Szabó – László Bertalan – Norbert Barkóczi

Purposes of aerial surveys and ground mapping with UAVs could be diverse so it is important to clarify what kind of UAV system and settings are necessary to start the workflow. We have to answer the following questions in order to prepare the best equipment and accessories for the specific area and research:

- How large is the area of interest?
- What should be the flying altitude for the appropriate coverage?
- What kind of product is needed? What specific kind of camera or sensor is essential for that survey?
- What specific kind of camera system (i.e., auto shutter) or camera mounting accessory (i.e., gimbal) is the most appropriate?
- What takeoff and landing conditions are to be expected in the operation area?

Unlike traditional manned aircraft, UAV technology is generally a low-cost solution for mapping purposes. Nevertheless, a complicated terrain configuration or specific types of survey require additional accessories that could increase the costs. The main component is the flying vehicle itself, but there are numerous additional components (e.g., telemetry, controllers, etc.).

Motors and propellers are usually powered by so-called lithium-polymer (Li-Po) batteries. The flight controller unit (FCU) handles speed and maneuvering. The FCU accepts input commands from the remote pilot or follows an autonomous flight program. The inertial measurement unit (IMU) records the actual attitude, accelerations, and altitude of the UAV. The FCU is directly connected to a (possibly differential) GPS/GNSS receiver and a solid-state compass (so-called magnetometer) for navigation. The latter assists in additional but useful functions like position-holding, "home-coming" automated procedures or the most important "navigation-through-waypoint" option. The camera mount system may include components for different types of controlled camera motion and image capturing defined by the ground operator. A "gimbal" is a three-axis stabilized mount that automatically holds the camera/sensors in the intended orientation. During a UAV flight, the flying platform and obviously the gimbal are subject to changes in attitude due to flight dynamics, turbulence, and wind. Therefore, the gimbal system adjusts the sensor orientation angle in all directions (up/down, forward/backward, left/right), often in a smooth way, thanks to a gyrostabilized mechanism. Datalink to and from the drone is usually set around the 2.4 GHhz frequency (Colomina and Molina 2014). The ground operator uses, normally, a remote controller with an onboard screen or an FPV (first-person view)/virtual reality goggle.

Typically, the concept of UAV-based aerial mapping is to perform several repeated flights over the area of interest and take images. These are later input to photogrammetric processing, in order to develop orthoimages and tridimensional models. Contrary to the classical stereophotogrammetry concept, nowadays, the so-called "structure-from-motion (SfM) method is used in general UAV-based mapping, for defining aerial image orientation and the orientation parameters. The advantage of SfM is that we are still able to reconstruct the three-dimensional geometry of an object by using multiple images taken from several viewpoints, without capturing location and orientation data of separate images (Fonstad et al. 2013). Compared to the aforementioned stereophotogrammetry, which relies on segments of overlapping images in necessarily parallel flight paths, SfM allows to determine X, Y, Z spatial coordinates of points of the target object from almost randomly captured imagery as long as the overlap among them is sufficient. The SfM approach operates without problems in case of oblique imagery as well.

UAV flight campaigns generally follow the workflow shown in Fig. 4.5. Each UAV-based aerial survey must be well prepared as far as both the analytical and operational parts are concerned. Analytical part involves determination of flight parameters and path over the area of interest, camera parameters, internal/exterior orientation, and the expected ground sampling distance (GSD) (Miřijovský and Langhammer 2015).

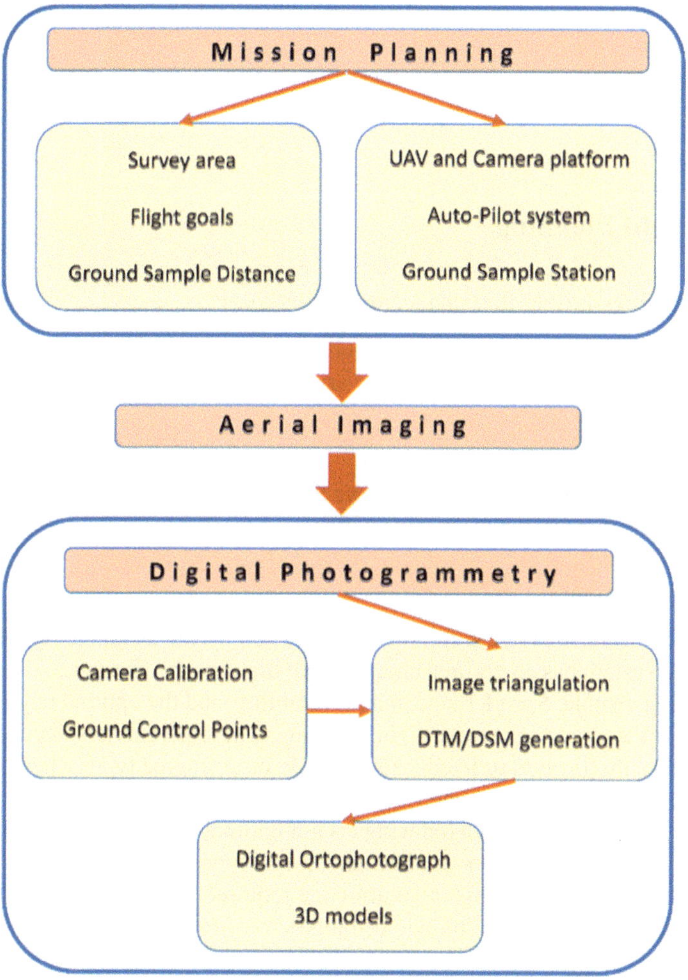

Fig. 4.5 General workflow of a UAV mapping procedure (Adapted from Nex and Remondino 2014)

4.3 Photogrammetry-Based Processing of Small-Format Aerial Images

Gergely Szabó – László Bertalan – Norbert Barkóczi

The basics and the concept of small-format UAS photogrammetry is more or less the same as the classic stereo image processing. The main difference (at least today) is related to the quality of the cameras. Classic photogrammetry cameras are extremely high-precision equipment. Cameras typically deployed on UAVs are low-cost instruments, with limited precision and resolution. In addition, in a UAV survey, the number of acquired images is (usually) large compared to a manned flight with metric cameras. The exterior orientation procedure of each image would take too much time, and the lower-quality, non-calibrated cameras would provide too large biases and too low accuracy. Therefore, recent UAV-based surveys focus on block formation and image triangulation using numerous tie points and few ground control points.

We are now about to delineate the main steps of a typical UAV-based image processing.

4.4 Flight Planning

Gergely Szabó – László Bertalan – Norbert Barkóczi

Before the flight, we have to plan the flight depending on the available camera and processing software. Some of the main variables are:

1. Area extent
2. Expected resolution (on the ground)
3. Camera field of view (FOV) to achieve the desired overlap
4. Specific features of the processing software
5. Survey goals
6. Legal background

The size of the area (1) is significantly affected by the available flying vehicle. Each UAV system has an optimal area size that matches the features and limitations of the flying vehicle. Speed, range, maneuverability, and the optimal payload of the vehicle are essential for correct flight planning. These parameters depend on type of UAV (i.e., multirotors or fixed-wing aircraft, etc.), engine type (whether electric or internal combustion), planned flight distance, weight of the camera, etc. For example, a fast vehicle with fixed wings is not optimal in a small, but detailed, area (e.g., 300×300 m built-up area), because in this type of mission, small base distances among images and high resolution are required. On the other hand, a rotary wing platform is not optimal for a long-range survey (e.g., river survey over several kilometers).

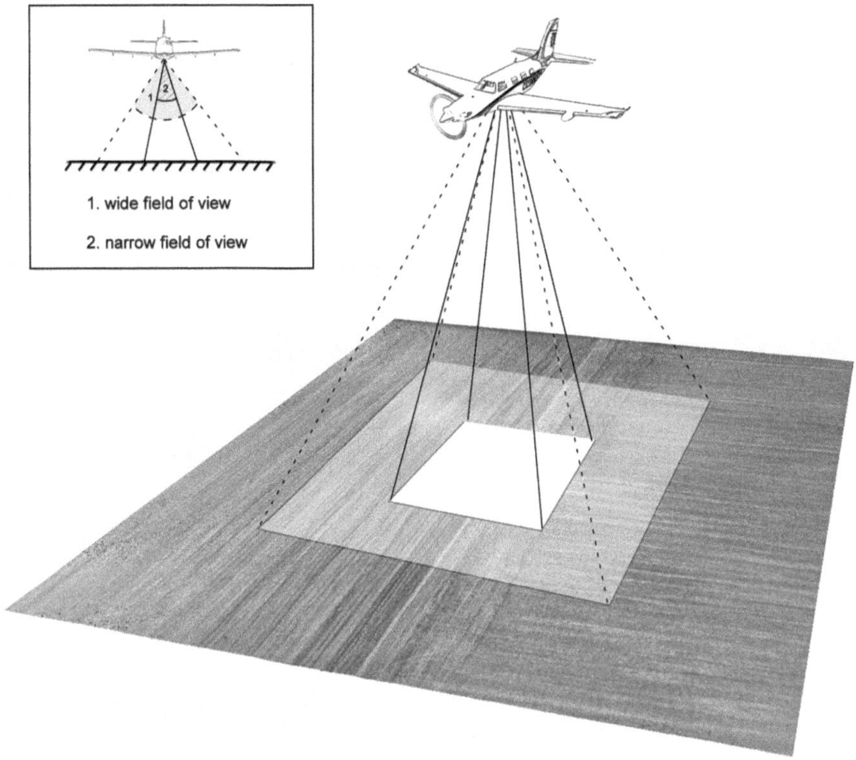

Fig. 4.6 Different types of field of view

Ground resolution (2) is the function of sensor resolution, type and quality of the lens, and flying height (i.e., the camera distance from the surface). The GSD (ground sample distance) is the distance between centers of two adjacent pixels. Features of the lens (3) define the field of view (FOV), which affect GSD and the ground coverage (Fig. 4.6).

The features of the processing software (4) are also important for flight parameters. Some software is optimized for processing row-flight image series only. The other is able to process nonrow-ordered images as well.

We need to know the purpose of the survey (5); a densely built urban environment requires a different kind of flight (e.g., smaller imaging base distance, shorter row spacing) than expected in surveying pasture or arable land (Fig. 4.7).

For intensively built urban areas, if the imaging base distance is large, the building tops "lean away", i.e., bases and roofs are not in orthogonal position (even on the orthophoto). This is known as radial distortion (Fig. 4.8) which increases proportionally to the distance from the center point (principal point).

Nowadays, legal conditions for the use of the drone (6) are slowly beginning to be clarified (in most countries). Regulation governs drone flights in urban and rural

Fig. 4.7 Aerial photos of a densely built area (*left*) and arable land (*right*)

Fig. 4.8 Orthophoto from few largely spaced images with radial distortion (*left*) and true ortho image from a high number of well-overlapping images (*right*)

environments, at public events, etc. However, these rules do not always allow for optimal survey conditions (e.g., closed airspace, flight limitations over crowd or nature reserve, etc.).

As we can see, flight planning is a complex process, with a theoretical "decision tree," where we can think over all elements of the workflow. For this reason, useful system applications have been developed that can automatically drive flights once the required parameters are specified by the user. Such is the case, for instance, of Pix4D Capture, DJI Go4, etc.

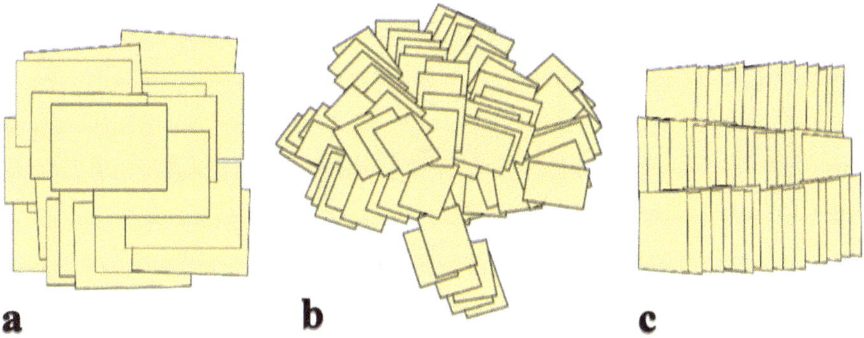

a b c

Fig. 4.9 Effects of different image acquisition modes on image block's quality: (**a**) manual mode with a scheduled interval, (**b**) low-cost navigation system with irregular image overlap, and (**c**) autopilot mode with a high-quality navigation system

The process of flight planning requires the above described information about the area of interest, the planned ground sample distance (GSD) and the interior orientation parameters of the mounted camera. Normally, the desired image scale and the focal length of the camera are fixed to derive the mission flying height. The waypoints of the path are technically the camera's perspective centers. They are to be computed with the purpose of ensuring the approximately 60–80% of longitudinal and transversal overlap of image strips (Nex and Remondino 2014). Geoscientific investigations in general need high overlaps and low-altitude flights in order to provide small GSDs. On the other hand, emergency surveying requires the observation of wider areas in a short time period, generally yielding a lower resolution.

Once the survey specifications, the aerial platform type, and the environmental conditions are selected, the UAV flight can be conducted in a manual, assisted, or autonomous mode.

The built-in GPS/GNSS navigation devices usually serve the autonomous flight procedures (takeoff, navigation, and landing). The performed flight typology has a great influence on the image network quality (Fig. 4.9). Manual mode provides usually very irregular image overlap and image acquisition geometry; these problems can be solved by using GPS-/GNSS-automated navigation. On the other hand, for very close-range acquisition over particularly rugged/complex terrain, manual mode may become necessary for coverage or obstacle avoidance reasons.

The navigation system or the so-called autopilot includes hardware and software as well. The autopilot performs the flight according to the planned route. The progress of the flight can be generally observed by the ground control station (GCS). As it is the case in manual mode flight, during automated operation a real-time flight data telemetry reports position, speed, altitude, distances, GPS/GNSS observations, battery charge status, rotor speed, etc.) (Fig. 4.10).

Geoscientific investigations may involve situations where due to the extremely specific terrain conditions the flight mission planning could not provide the acceptable coverage results. Manual flying is then necessary. The rhyolite tuff badlands in Hungary is a good example: bare tuff surfaces are intermingled with forest patches

Fig. 4.10 Example of a flight planning interface for UAV mapping (Image: GREAL – European University of Rome)

and the elevation differences are impressively high. The shape of the badland valley can be overly complicated for flight planning since it requires several adjustments of flying heights. In this case, the remote pilot had the responsibility to provide efficient accuracy and coverage by capturing the images from two height levels, taking into account the approximate slope gradient of the valley surface (Fig. 4.11).

Whether flight planning/autopilot software is used or not, we need to consider the number of aerial photos and their locations. The more images are acquired, the more hardware resource we need to process them. It is therefore recommended to use the least but sufficient number of images considering the 80% image overlapping (Figs. 4.12 and 4.13).

In addition to the number of images (from a different point of view, image density), the perspective distortion caused by the inclination of images needs to be mentioned as well. For gimbal-free UAV systems, the camera loses its nadir-looking position relative to the surface (in other words, the axis of the camera is not aligned as perpendicular to the surface as possible), because the flying vehicle rotates around its three axes (yaw, pitch, roll) so that perspective distortion occurs (Fig. 4.14).

Apart from the perspective distortion, the disadvantage of oblique images is that surface resolution cannot be fully controlled. The resolution of the distant areas will be smaller, causing inhomogeneity (Figs. 4.15 and 4.16).

Fig. 4.11 Path of the multilevel flight above the study area

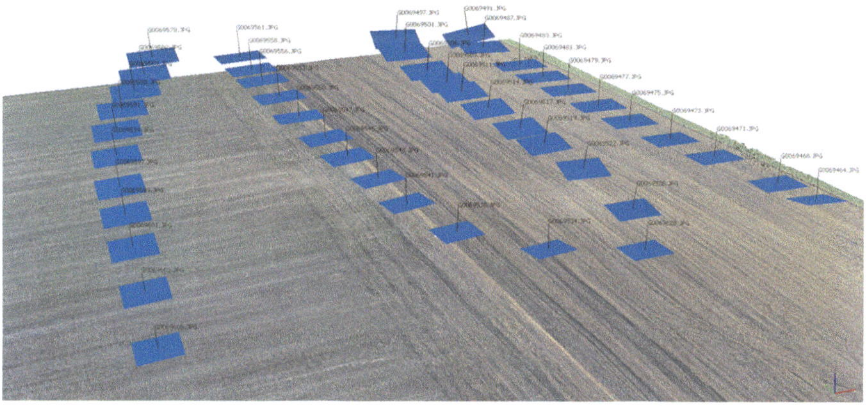

Fig. 4.12 Imaging in nearly parallel rows

4.4.1 Influencing Factors of Drone-Based Imaging

The result of photogrammetry-based image processing could be affected by several factors. Surface pattern plays an evident key role in the progress; however, other influencing variables have to be taken into account.

The most important parameters and factors are the following:

- Type of surface pattern/homogeneity (i.e., water, snow, concrete, etc.)
- Image sharpness
- Blur exemption

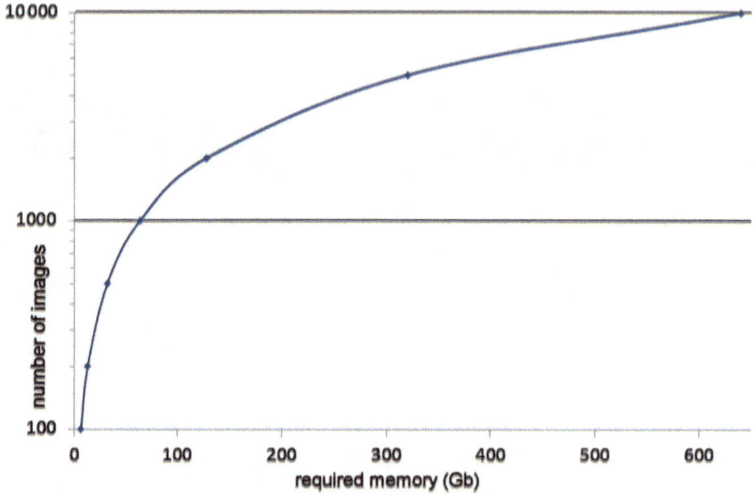

Fig. 4.13 The required system memory depending on the number of images in Agisoft PhotoScan (Based on Agisoft 2017)

Fig. 4.14 Images that are not parallel to the surface

- Transparent and glittery areas and antisolar halo
- High-detailed objects (e.g., canopy) compared to the given resolution
- Regularly repeating pattern (i.e., arable lands, built-up areas)
- Object in motion (e.g., vehicles)
- Exposure characteristics (e.g., shutter)
- Compression details of the digital imagery
- Quality of optical sensor
- Light conditions
- Others

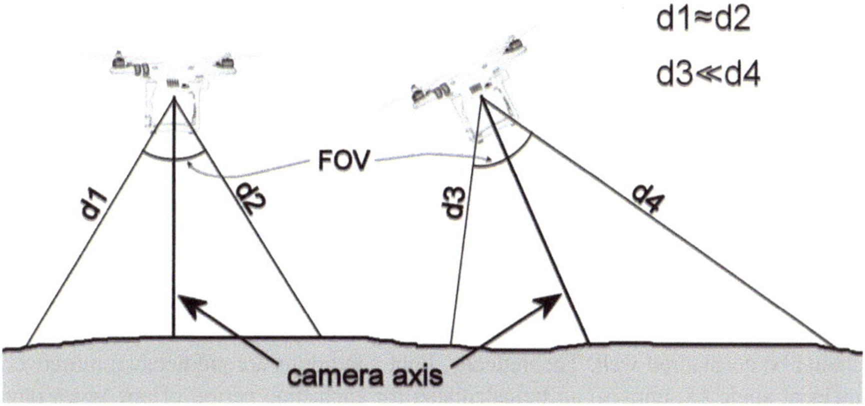

Fig. 4.15 Perpendicular (*left*) and oblique camera axis (*right*)

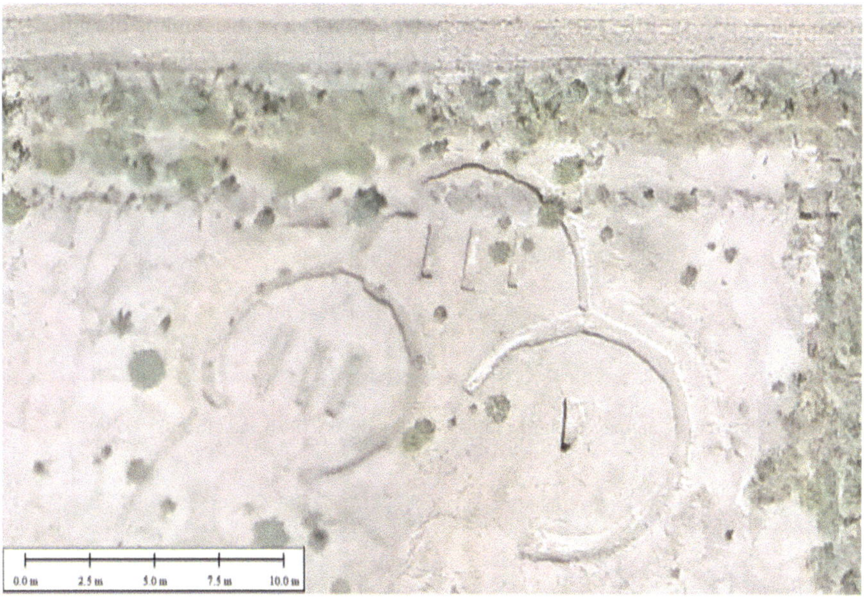

Fig. 4.16 Altering resolution on an orthoimage

Sharpness, blur, compression parameters, and the exposure characteristics are more or less predictable and can be managed by appropriate settings. However, it is clear that some trade-off may always be necessary. If the shutter speed is fast, then the image might turn out to be underexposed. In order to avoid this, sensitivity (i.e., ISO value) should be increased, which, in turn, increases the image noise. All aspects must be properly budgeted.

Optical and capturing properties of the camera are also part of this topic. Neither a device with a low-quality sensor nor a low-resolution camera lens (with low LW/

PW values) is able to capture adequate images and produce good-quality material. Exposure type also affects the image quality: different distortions and disturbances occur in case of low light conditions, depending on whether the camera has a mechanical shutter, central shutter, or no mechanical shutter at all.

Another important aspect is how the camera builds up the image from raw pixels; namely, the recent and popular RGB filter (so-called Bayer filter) systems do not forward the colored image directly; rather, they "mix" it through a complex method performed by the camera processor (Enyedi et al. 2016).

Another question is the compression method used by the processor; moreover, there is the possibility to parameterize or omit the whole compression phase.

In order to have appropriate contrast and brightness, date and time of the imaging should be considered well. Theoretically, light conditions are predictable; moreover; incident angle of sunrays can be calculated for each time period of any given day; however, luminous transmittance of air (e.g., due to clouds, fog, dust, etc.) significantly affects the local light conditions and consequently the contrast as well (Fig. 4.17).

Shining and antisolar halo depend upon the angle of incidence of sunrays and cannot be filtered in all occasion. Therefore, both phenomena have a remarkable influence on image quality as they alter values of involved pixels, image by image (Fig. 4.18).

Fig. 4.17 Differences in image contrast in case of cloudy sky (*left*) and clear sky (*right*)

Fig. 4.18 Antisolar halo (*left*) and glittering (*right*) on water surface

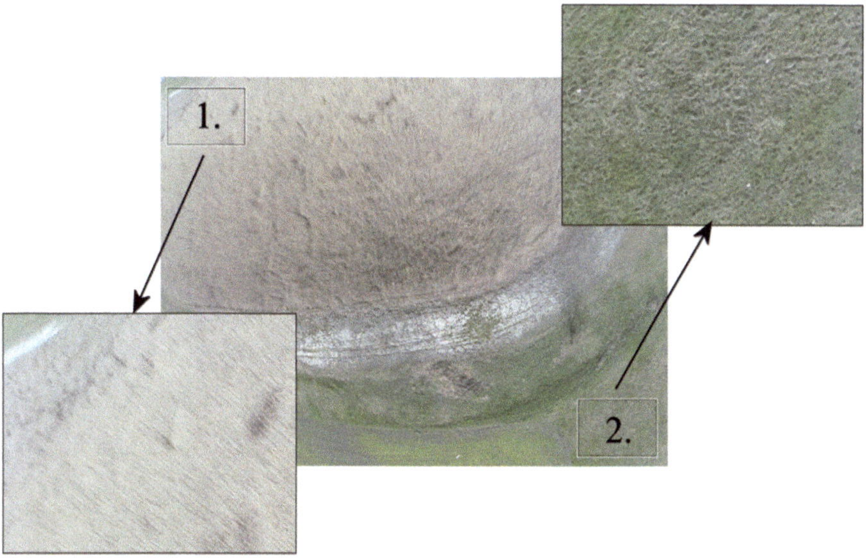

Fig. 4.19 Effect of compression in case of different patterns

High-detailed patterns of objects within an image, in reference to a given resolution, may cause problems not only during image processing, but even during image capturing as well. Figure 4.19 represents imagery taken at Hortobágy (Hungary). As a result of compression, reeds at the top left part of the area are blurred (1), while pasture at the bottom right corner (2) is sharp.

This problem originates from the compression algorithm applied by the camera processor: in some cases, it can significantly degrade the resolution within the pattern field.

The issue of overly detailed objects arises during the photogrammetric processing as well. Then, small details become blurred, or the software is unable to identify all details on each image during the processing. Typical examples are trees and their canopy, which appear jagged even at extremely high resolution. Furthermore, such extremely complex structures may also suffer from slight motions due to wind. In this case the software calculates an averaged canopy "surface" to provide a generally acceptable model of it.

Nevertheless, during the leafless winter period, bare branches usually cause hardly resolvable anomalies for the processing software which frequently identifies the spatial location of these points incorrectly (Fig. 4.20).

Considering the current development of processing software, objects caught in motion through an image capture series do not significantly affect the whole image processing phase: they are automatically filtered by the softwares. Yet, these can be found in a duplicated (or even in a multi-duplicated) form on the end products (especially on the orthophoto) with a disturbing effect (especially for quantitative analysis) (Fig. 4.20).

Fig. 4.20 False identification of bare branches (*left*), multiplied objects that move during the aerial imaging (*left*)

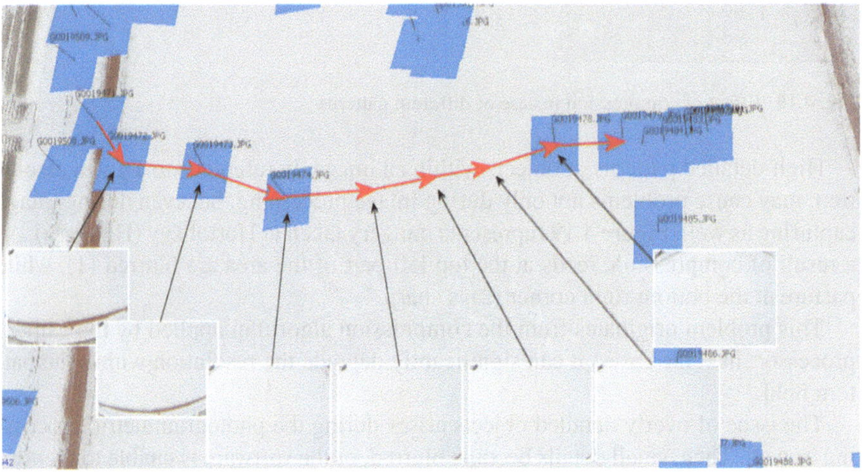

Fig. 4.21 Images omitted from alignment due to the lack of matching points (*red arrows* indicate the flight path)

When the target surfaces show too regular and homogeneous patterns, the image processing is difficult and lumbering even with correct exposure due to the lack of enough identifiable matching points. The processing software is not able to align the homogenous details, whether to the adjacent image or to a farther one (Fig. 4.21).

In the absence of matching points, the given image (or images) might be omitted from the processing, so the overall quality of the orthophotographs and the surface model could be compromised. If a very regular surface is to be surveyed, an artificial matching point network could be established before the flight (Fig. 4.22).

The issue of comprehensive photogrammetric surface modeling of water is still not solved yet. It is generally impossible as a result of waving motions and shiny surfaces. In this case, a perfectly flat and horizontal surface can be assumed, and then the incorrectly calculated surface points can be removed during post-processing (Fig. 4.23).

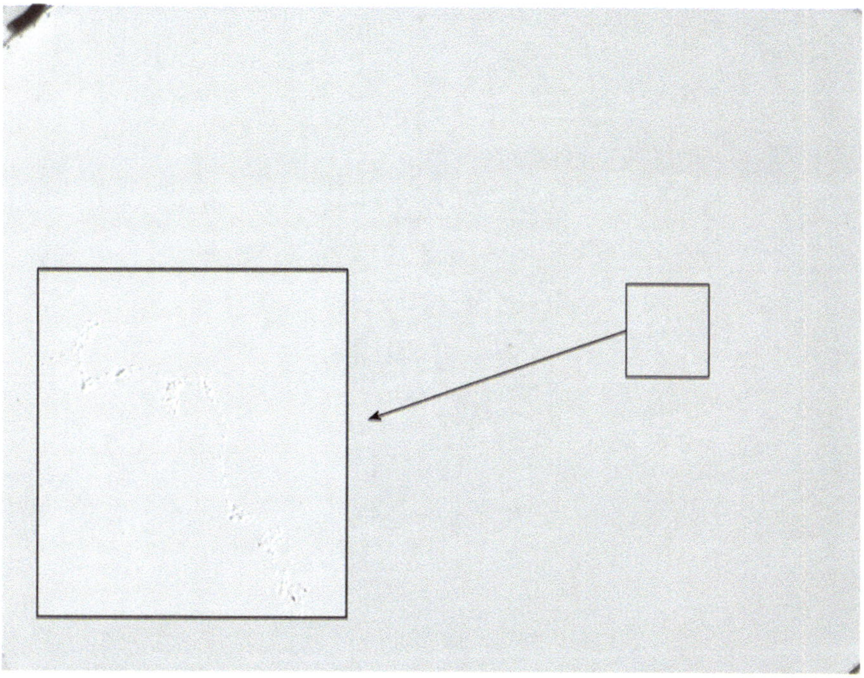

Fig. 4.22 Establishment of matching points on a homogenous surface

Fig. 4.23 Incorrectly calculated surface in case of stream water

Fig. 4.24 Birds, between the UAV and the surface

Finally, several other factors may interfere with the imaging. For example, birds often fly close to the vehicle (Fig. 4.24), generating false (and altering) points.

4.5 Image Processing

Gergely Szabó – László Bertalan – Norbert Barkóczi

The image processing of close-range photogrammetry basically consists of the same steps as the traditional photogrammetry. Once images are oriented and calibrated, surface model and orthoimage extraction can be automated (Colomina and Molina 2014). However, there are differences between the two methods.

4.5.1 Calculate Lens Distortion

Every type of camera lens (not only the types used in aerial imaging) distorts the captured image to some extent. In spite of the excellent optics of traditional metric aerial cameras, the quality of the UAV cameras is generally worse: knowing

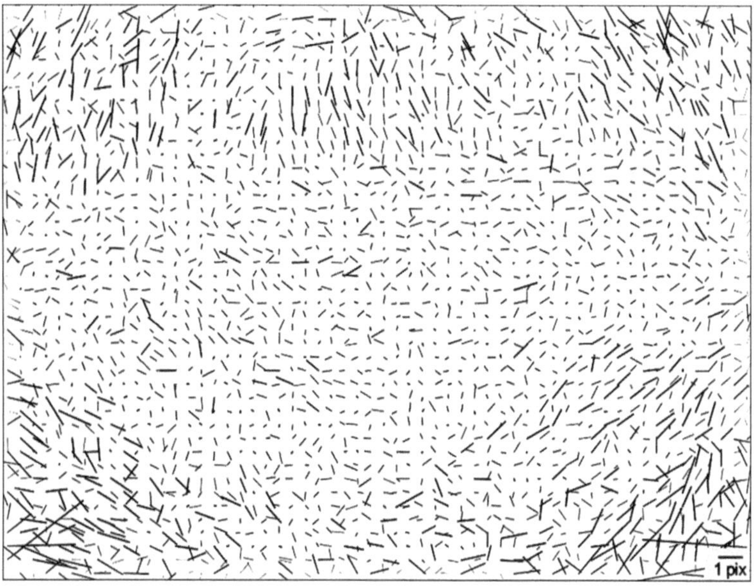

Fig. 4.25 Images residuals of an action camera (Agisoft PhotoScan)

properties and parameters of the optics is essential for further processing. In fact, the removal of lens distortion from an image is a significant part of UAV photogrammetry.

Most software packages handle lens calibration along with image orientation, calibration, and alignment (and aerial triangulation as well). Specific software packages are designed to perform high-quality aerial triangulation (e.g., BINGO). Generally, however, 3D software (e.g., Agisoft PhotoScan, Pix4D, etc.) has built-in capability to this process (Fig. 4.25). These products examine the optical system of the camera and calculate the systematic image deformations automatically or with additional extensions/plugins (e.g., Agisoft Lens).

Sharp images and homogenous optical parameters are essential during the flight. It is then advisable to avoid excessively complex, zoom lens systems, because vibration may change the distortion parameters during the flight.

4.5.2 *Image Alignment: Aerial Triangulation, Block Adjustment*

In traditional photogrammetry, photo alignment was executed by placing marks to points of interests. These points were identifiable on the stereopair images, and by using them (based on parallaxes), the analyst could calculate the relative spatial positions of images. Nowadays, high-capacity computers calculate many more

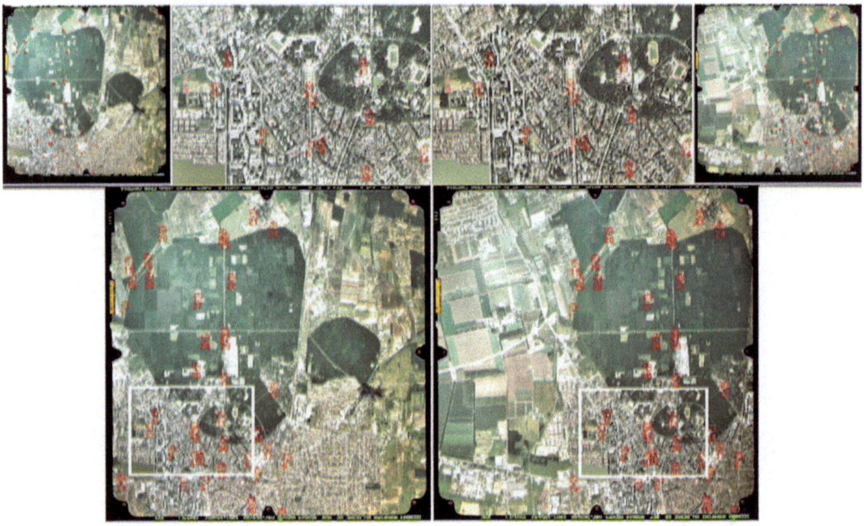

Fig. 4.26 Tie points on two images

marks (even thousands), so-called tie points to align images to each other. During this procedure, the software localizes tie points on images by testing their neighborhood, with the purpose of finding the best matching (Fig. 4.26).

In most current software, tie-point detection is embedded into a complex process, which includes interior orientation, exterior orientation, and aerial triangulation.

In classic interior orientation, the analyst had to place the negative image into a local coordinate system and define the fiducials on aerial images. Besides these points, the knowledge of camera focus was essential as well. Now it is possible to define these data, but in the lack of these information (e.g., in case of a simple digital camera), the process is automatic, and the software acquires the necessary information during the calculation (Jancsó 2010).

The aim of exterior orientation is to locate the processed model in a reference system (i.e., projection system). For this purpose, we look for points that can be clearly identified in the images. The already mentioned tie points are also suitable as control points if they are localizable on the images. Thus, the ground control points (GCPs) are independent from the tie points, but a tie point with known external coordinates is suitable to play both roles (Fig. 4.27).

Nowadays, in most software, alignment is just one part of a complex process. Using a series of overlapping images, the software creates a block to perform aerial triangulation (Jancsó 2010). The aim of triangulation is to extract exterior orientation parameters in relation with image position. Since images were taken from different positions, we can connect each common point on overlapping images with one ray (image ray). During air triangulation the software eliminates errors, using image rays (block adjustment).

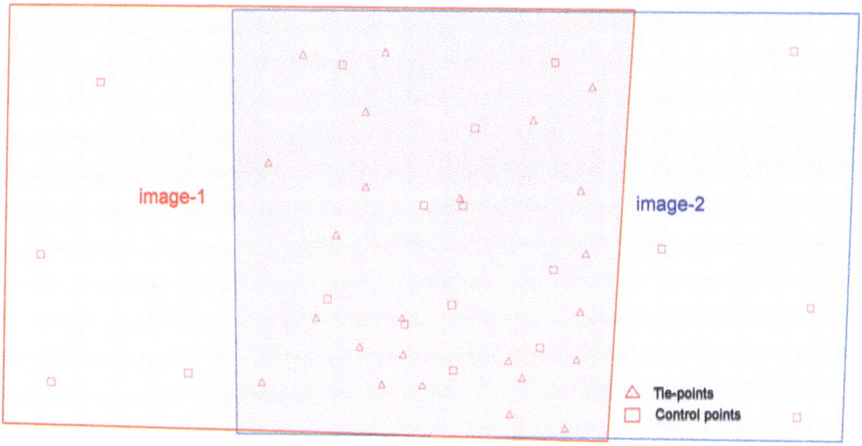

Fig. 4.27 Tie points and control points

In some modern software, the user has limited possibilities to parameterize the process. For example, with Agisoft PhotoScan the user can set some basic parameters for image alignment, such as the maximum number of switch points, the number of pixels used from photos, etc. After alignment, images are connected to each other (if the process was successful). Defining the exterior orientation is possible in two ways.

If we have recorded the position of the images (1) and the three rotation angles (yaw, pitch, roll) during the flight, these data can be loaded into the software. It is necessary to know these data with high precision if we want to get an exact 3D model and/or orthoimage. Recently, differential GPS and IMU (inertial measurement unit) data became available for generating highly precise positional and attitude data. If the accuracy of the orthophoto or surface model is not required at geodetic level, the data recorded by UAV's onboard standard GPS and IMU systems may be sufficient as well.

Another possible solution is to place GCP targets on the ground before flying. The GCPs are pre-marked artificial features that will appear on the aerial photographs, and their locations are known in relation to a reference system. The overall accuracy of GCP coordinate measurement should be within ±10 mm error (d'Oleire-Oltmanns et al. 2012). GCP markers can be different depending on application; their size and material vary in a wide range from round plastic shapes to square paper targets. In case of difficult terrain conditions, if there is no possibility to install permanent points, alternative solutions can be chosen, e.g., airbrush-based cross hairs on steep rock surfaces (Fig. 4.28), although there are some software products for which certain pattern is to be used for automatic recognition.

The number of GCPs is variable (at least three are required); basically, the more points we install, the more stable the exterior orientation will be. In practice, it may happen that some of the GCPs end up being not included in the images, or they are incorrectly identified, or their position was inaccurately measured in the first place

Fig. 4.28 Different types of GCP markers

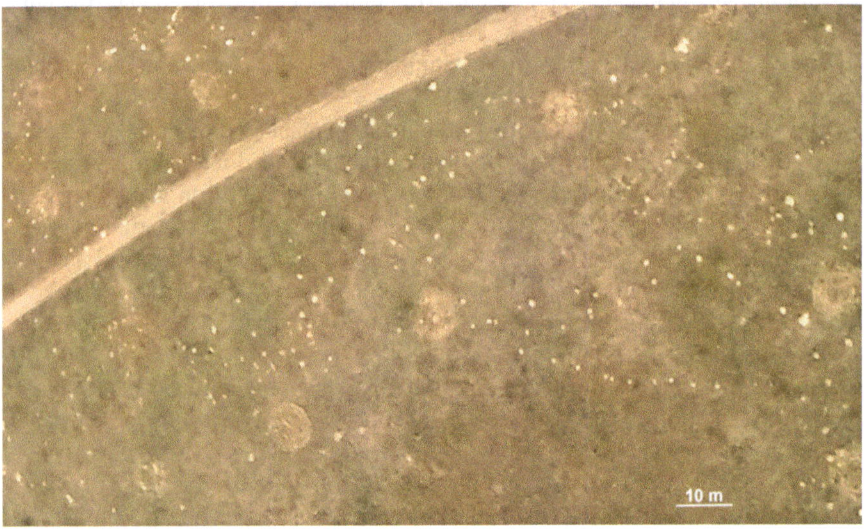

Fig. 4.29 Example of unidentifiable GCP. Among the molehills in the image, it is almost impossible to localize a white, round control point marker

(Fig. 4.29). In these cases, they should be omitted from processing and software packages allow to do so.

Spatial distribution of GCPs over the area of interest has a significant effect on automatic aerial triangulation. Traditional aerial photogrammetry required only four or five symmetrically distributed points, but for the recent structure-from-motion algorithms, the entire area should be covered by properly installed GCPs. In

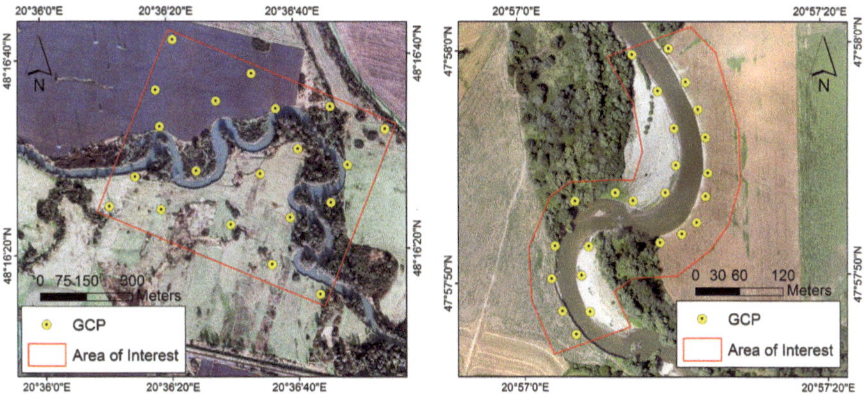

Fig. 4.30 Different spatial patterns for GCP distribution

order to provide an overall acceptable accuracy of digital orthophotograph or surface model, then the GCP structure is a crucial step of field preparation. According to several studies and statistical approaches, it had been stated that the GCP distribution should avoid linear patterns. On natural and plain terrains, a matrix should be created to build a nonlinear distribution of control points. In several cases, however, this is not possible. In geoscientific investigations, especially in geomorphology, various linear landforms exist. If the purpose is to monitor the planform changes of these landforms, then GCPs have to correspond to the observed spatial structure, to ensure accuracy in those areas. Figure 4.30 represents both possible situations. In the example on the left, the purpose was to create a digital orthophotograph of a selected area along the Hungarian Sajó River; in the example on the right, the main aim was monitoring of intensive lateral bank erosion processes of the meandering Sajó River.

Theoretically, for an approximately 1–2 km^2 large area of interest, 20–50 GCPs are necessary on the field in order to provide a high-accuracy survey. Consequently, the GCP installation and survey are a time-consuming part of the UAV mapping workflow. At an area of interest with a size of 1–2 km^2, this work stage could last around 2–3 h or maybe more, if difficult terrain conditions are present. A typical case is when GCPs are installed on both sides of a river channel. If there is no bridge or ferry, it may be necessary to perform several travels or to split work groups. Nevertheless, the duration of GCP setup can be decreased by the establishment of permanent control points at the area if the investigation will be performed repeatedly. Unfortunately, this is not always possible. For instance, doing so in agricultural parcels may cause disruption to local activities.

Using GCPs, the software generates equations by which each point in the model gets real external coordinates. For each GCP we get a root-mean-square error (RMSE) that specifies difference among the original positions of the GCP and the calculated, equation-based position of the same GCP. It may be more effective to omit the GCPs with too high RMSE values, in order to increase the overall accuracy. It may also happen that increasing the number of GCPs does not improve the

Fig. 4.31 Dense point cloud

accuracy. This can be caused, for example, by an incorrectly surveyed GCP. These points should also be omitted.

After GCP localization in a few images, most software products calculate an estimated GCP position on the other images as well.

Using all the calculated and defined data, the software calculates the depth information of the cameras, and a dense point cloud is created (Fig. 4.31).

The density of the point cloud depends on the user's input parameters, but there is a certain limit that is not advisable to cross. Generally, the number of images, their quality, and the processing methods determine the practical density of point cloud. The dense point cloud gained from photogrammetry is similar to the LiDAR-based point cloud. The greatest difference is that modern LiDAR datasets are multilayer (due to multiple reflection), while photogrammetry-based point clouds only return the topmost level.

Connecting adjacent points of the cloud gives a polygonal surface. A polygonal mesh model can then be constructed, to yield a surface. From this, most software generate and export a raster surface model (Fig. 4.32).

4.5.3 Working with Point Clouds

As previously mentioned, point clouds can be used to model surface and objects located on the surface as well. The photogrammetry-based point cloud (as opposed to LiDAR technique) is only suitable for calculating the topmost surface (digital surface model, or DSM) which refers to the real surface, with all features located on the ground (Fig. 4.33). In many cases DSM is the appropriate database, for example,

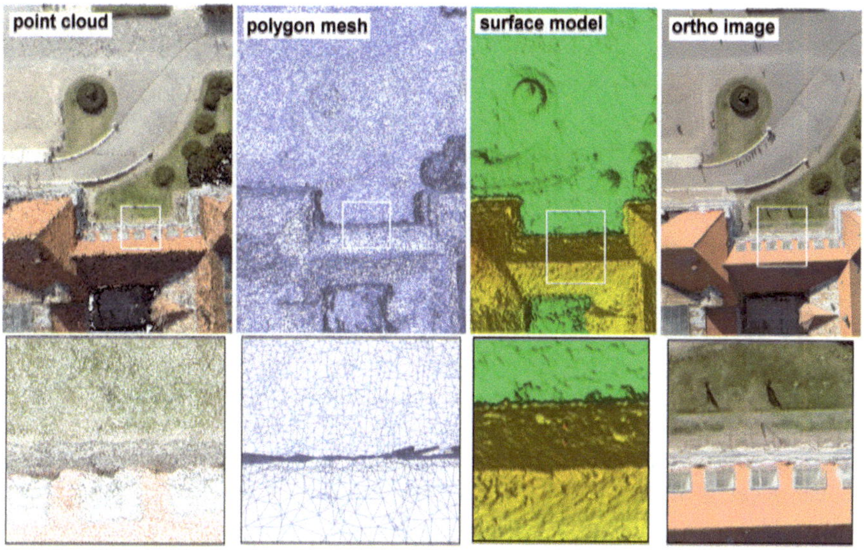

Fig. 4.32 From point cloud to orthoimage

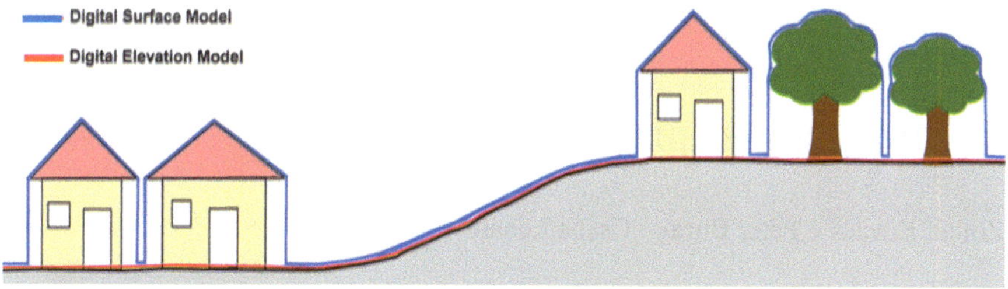

Fig. 4.33 Digital surface model and digital elevation model

to examine viewshed from a specific location, modeling wind channel effect among buildings, or calculate the potential solar energy of the roofs (Szabó et al. 2015).

In other cases, a feature-free model is needed: this is the case, for instance, in hydrological calculations or in several types of mapping. The digital elevation model (DEM) refers to the elevation of the original terrain, without landcover and artificial objects (i.e., vegetation, buildings, etc.).

Extracting a DSM from an image-based survey is not complicated, but it is also possible to generate a DEM as well, although its detail is currently below that achievable by LiDAR. A widespread processing software, i.e., Agisoft PhotoScan, can be used to define a cell size within which the lowest point is searched as the potential surface. In the next step, the software examines how this point relates to its own environment with regard to height differences and the angle of inclination (Agisoft). The result is a DEM, without surface features. From the location of the objects, the software removes the points, so the obtained surface is the result of the

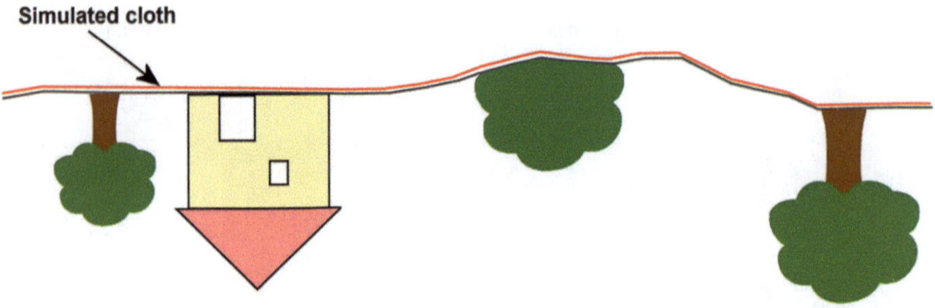

Fig. 4.34 The cloth simulation filtering (CSF) method

interpolation from the boundaries of the removed objects. Another software, CloudCompare, uses the cloth simulation filtering method (Zhang et al. 2016). The software reverses the model with objects pointing downward. Then it places a theoretical sheet ("cloth") on the reverse surface. This is defined as "true" relief (Fig. 4.34).

The filter separates the points of the ground and the objects. A good example of using these methods is a forest area (Fig. 4.35). The forest is suitable for this procedure, even in leafy conditions, if there are smaller or larger areas with no trees. During foliage-free period, the process is simpler.

4.6 Drone-Based LiDAR Technology

Zoltán Kovács – Péter Burai – Csaba Lénárt

4.6.1 *UAV-Based Laser Scanning (ULS)*

Until recently, UAVs were used mainly for photogrammetric purposes. Due to the dynamic development of UAV and LiDAR technology in the last years, UAS with higher payload capacity and longer endurance, equipped with new, commercially available compact LiDAR sensors is capable of acquiring 3D point clouds for civil applications including archaeology, open-pit mining, geography, hydrology, and forestry. The lower operating altitude of UAV-borne laser scanning, along with the ability to hover above a point or the possibility of setting arbitrary flying speed, results obviously in higher point density compared to the traditional airborne measurements. On the one hand, the lower operating altitude results in detailed acquired data (especially in case of full-waveform LiDAR) due to the smaller laser footprint diameter. On the other hand, due to the low operation time of battery-powered UAVs, the size of measurable investigation area is significantly decreased compared to traditional airborne solutions.

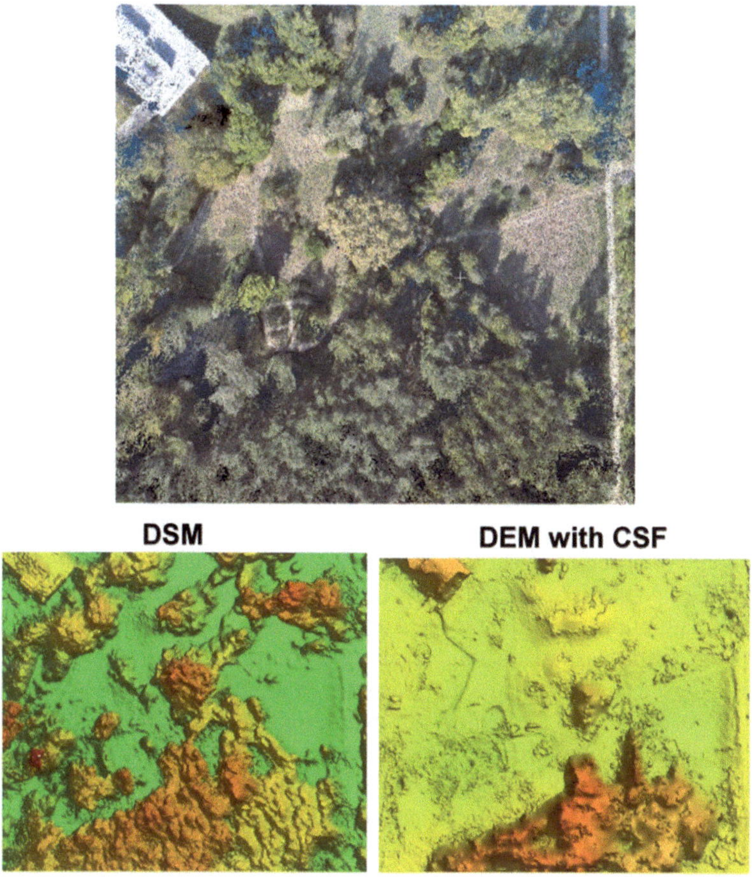

Fig. 4.35 DSM (*left*) and DEM derived from the point cloud by CSF method (*right*)

Unlike photogrammetry, LiDAR is able to see through vegetation canopies and provides data about the terrain. It proves more useful in modeling narrow objects such as transmission lines, pipes, and sharp-edged features. Detailed elevation data are mandatory to simulate flood events, and more accurate 3D shapes of the landscapes contribute to a deeper understanding of the geomorphological processes. Laser scanning in open-pit mining areas is an efficient and cost-effective way to collect data for volume calculation and the analysis of stockpiles and sediments (Höfle and Rutzinger 2011). LiDAR technology is capable of capturing complex objects – in excavations sites or buildings – with high resolution and high accuracy. Furthermore, ULS can be used without any direct physical access to the surveyed area, so it is safe in archaeology and cultural heritage applications (Amon et al. 2015). Full-waveform scanners, in particular, allow for extraordinary insight into our environment, e.g., forest structures, by sensing a vertical profile from canopy to ground (Pfeifer et al. 2015).

There are many small compact LiDAR sensors available for UAVs. Riegl VUX-1UAV LiDAR sensor (3.5 kg, ~7.7 lbs) has 10 mm survey-grade accuracy with measurement rate up to 500,000 measurements per second and 330° field of view (FOV). Velodyne LiDAR's HDL-32E sensor (2 kg, ~4.4 lbs) is able to make 700,000 points per second within 80–100 m (~262 to ~328 ft) range, with 360° horizontal FOV and +10° to −30° vertical FOV.

4.6.2 Airborne Laser Bathymetry (ALB)

The topobathymetric LiDAR is able to capture very high-resolution topographical information for shallow water areas and the surrounding littoral zones simultaneously. Measuring bathymetry by NIR laser is not possible due to the reflection and absorption of the laser light at the water surface. Traditionally, ship equipped with a multibeam echo sounder (MBES) system is able to measure high-resolution bathymetry, but it does not cover the bathymetry in the shallow water because of ship draft limitation (Andersen et al. 2017). The introduction of scanners emitting short and narrow laser pulses (pulse duration <=1.5 ns, beam divergence ca. 1 mrad) in the green domain of the spectrum ($\lambda = 532$ nm) changed the scenario. This wavelength is found to be the least attenuating, resulting in the largest penetration depth for the laser (Andersen et al. 2017) at a high scan rate of more than 200 kHz (Mandlburger et al. 2015). Separating the returns of air-water interface and the water-body floor makes it possible to map smaller inland waters (e.g., mountain streams, medium-size rivers). The penetration depth in water is limited by the attenuation of the laser beam, and one of the biggest operational constraints for topobathymetry LiDAR is water clarity. Submerged LiDAR points have to be corrected because of the refraction of the laser beam at the air-water interface according to Snell's law. Green and NIR LiDAR are often used simultaneously to get returns from the seabed and the interface at once. Suspended sediment, water molecules, and dissolved material all act on the laser beam by scattering and absorption. The consequence is a substantial reduction in power as the signal propagates into the water (Guenther 2011).

At the lower course of the pre-alpine Pielach River near Neubacher (Austria) ALS, ULS, and ALB measurements were compared by Mandlburger et al. ALS data was collected by an Riegl LMS-Q 1560 dual-channel, full-waveform topographic laser scanning system installed onboard a Diamond DA 42 light aircraft flying at 600 m (~2,000 ft) above ground level. ULS data capturing was carried out with the Riegl VUX-SYS sensor system mounted on a RiCOPTER UAV platform, 50 m (~160 ft) above ground level. ALB survey was carried out with the Riegl VQ-880-G topobathymetric laser scanner whose flying altitude was 600 m. The point density was the most homogeneous in case of ALB due to its water penetration capability, while ALS and ULS provide water reflections around the nadir direction. The study demonstrates that ULS can complement ALB in alluvial environments due to the ALS and ULS having higher point density (ULS with overall 1,100 points/m2) (Mandlburger et al. 2015).

References

d'Oleire-Oltmanns S, Marzolff I et al (2012) Unmanned aerial vehicle (UAV) for monitoring soil erosion in Morocco. Remote Sens 4(11):3390–3416. doi:10.3390/rs4113390

Agisoft (2017) Dense cloud classification and DTM generation with Agisoft photoscan professional. Available via http://www.agisoft.com/index.php?id=35. Last accessed 8 May 2017

Amon P, Riegl U et al (2015) UAV-based laser scanning to meet special challenges in lidar surveying, Geomatics Indaba Proceedings 2015. Stream 2:138–147

Andersen MS, Gergely Á et al (2017) Processing and performance of topobathymetric lidar data for geomorphometric and morphological classification in a high-energy tidal environment. Hydrol Earth Syst Sci 21:43–63. doi:10.5194/hess-21-43-2017

Candiago S, Remondino F et al (2015) Evaluating multispectral images and vegetation indices for precision farming applications from UAV images. Remote Sens 7(4):4026–4047. doi:10.3390/rs70404026

Carrio A, Pestana J et al (2015) UBRISTES: UAV-based building rehabilitation with visible and thermal infrared remote sensing. In: Reis L, Moreira A, Lima P, Montano L, Muñoz-Martinez V (eds) Robot 2015: second Iberian robotics conference, Advances in Intelligent Systems and Computing, vol 417. Springer, Cham. doi:10.1007/978-3-319-27146-0_19

Colomina I, Molina P (2014) Unmanned aerial systems for photogrammetry and remote sensing: a review. ISPRS J Photogramm Remote Sens 92:79–97. doi:10.1016/j.isprsjprs.2014.02.013

Enyedi A, Kozma Bognár V, Berke J (2016) Távérzékelési célú képalkotó algoritmusok összehasonlítása tartalom és szerkezet alapján. Remote Sens 6(6):464–475

Essen H, Johannes W et al (2012) High resolution W-band UAV SAR. Paper presented at the 2012 IEEE International Geoscience and Remote Sensing Symposium (IGARSS), 22–27 July 2012. doi: 10.1109/IGARSS.2012.6352480

Fonstad MA, Dietrich JT et al (2013) Topographic structure from motion: a new development in photogrammetric measurement. Earth Surf Process Landf 38(4):421–430. doi:10.1002/esp.3366

Georgopoulos A, Oikonomou C et al (2016) Evaluating unmanned aerial platforms for cultural heritage large scale mapping. Int Arch Photogramm Remote Sens Spat Inf Sci XLI-B5:355–362. doi:10.5194/isprsarchives-XLI-B5-355-2016

Grenzdörffer G, Niemeyer F, Schmidt F (2012) Development of four vision camera system for a micro-UAV. Int Arch Photogramm Remote Sens Spat Inf Sci XXXIX-B1:369–374

Guenther GC (2011) Airborne lidar bathymetry. In: Maune D (ed) Digital elevation model technologies and applications: the DEM users manual. Maryland, Asprs Publications, p. 253–320

Höfle B, Rutzinger M (2011) Topographic airborne LiDAR in geomorphology: a technological perspective. Z Geomorphol 55(2):1–29. doi:10.1127/0372-8854/2011/0055S2-0043

Jancsó T (2010) Data acquisition and integration 5., Photogrammetry. Available via http://www.tankonyvtar.hu/hu/tartalom/tamop425/0027_DAI5/ch01.html. Accessed 8 May 2017

Kalisperakis I, Stentoumis C et al (2015) Leaf area index estimation in vineyards from UAV hyperspectral data, 2D image mosaics and 3D canopy surface models. Int Arch Photogramme Remote Sens Spat Inf Sci XL-1/W4:299–303. doi:10.5194/isprsarchives-XL-1-W4-299-2015

Kohoutek T, Eisenbeiss H (2012) Processing of UAV based range imaging data to generate detailed elevation models of complex natural structures. Int Arch Photogramm Remote Sen Spat Inf Sci XXXIX-B1:405–410

Levin E, Zarnowski A, McCarty JL, Bialas J, Banaszek A, Banaszek S (2016) Feasibility study of inexpensive thermal sensors and small UAS deployment for living human detection in rescue missions application scenarios. Int Arch Photogramm Remote Sens Spat Inf Sci XLI-B8:99–103

Mandlburger G, Pfennigbauer M et al (2015) Complementing airborne laser bathymetry with UAV-based lidar for capturing alluvial landscapes. Proc. SPIE 9637. Remote Sens Agric Ecosyst Hydrol XVII:96370A. doi:10.1117/12.2194779

Miřijovský J, Langhammer J (2015) Multitemporal monitoring of the Morphodynamics of a mid-mountain stream using UAS photogrammetry. Remote Sens 7(7):8586–8609. doi:10.3390/rs70708586

Nebiker S, Lack N et al (2016) Light-weight multispectral UAV sensors and their capabilities for predicting grain yield and detecting plant diseases. Int Arch Photogramm Remote Sens Spat Inf Sci XLI-B1:963–970. doi:10.5194/isprsarchives-XLI-B1-963-2016

Nex F, Remondino F (2014) UAV for 3D mapping applications: a review. Appl Geomatics 6(1):1–15. doi:10.1007/s12518-013-0120-x

Pfeifer N, Mandlburger G et al (2015) Lidar: exploiting the versatility of a measurement principle in photogrammetry. In: 55th photogrammetric week 2015. Stuttgart, Germany, p 105–118

Remy M, de Macedo K, Moreira J (2012) The first UAV-based P- and X-band interferometric SAR system. Paper presented at the 2012 IEEE international geoscience and remote sensing symposium (IGARSS), 22–27 July 2012. doi:10.1109/IGARSS.2012.6352478

Rosen PA, Hensley S et al (2007) UAVSAR: new NASA airborne SAR system for research. IEEE Aerosp Electron Syst Mag 22(11):21–28. doi:10.1109/MAES.2007.4408523

Scholtz A, Kaschwich C et al (2011) Development of a new multi-purpose UAS for scientific application. Int Arch Photogramm Remote Sens Spat Inf Sci XXXVIII-1/C22:149–154

Szabó S, Enyedi P et al (2015) Automated registration of potential locations for solar energy production with light detection and ranging (LiDAR) and small format photogrammetry. J Clean Prod 112(5):3820–3829. doi:10.1016/j.jclepro.2015.07.117

Wallace L, Lucieer A et al (2012) Assessing the feasibility of UAV-based LiDAR for high resolution forest change detection. Int Arch Photogramm Remote Sens Spat Inf Sci XXXIX-B7:499–504

Xie F, Lin Z et al (2012) Study on construction of 3D building based on UAV images. Int Arch Photogramm Remote Sens Spat Inf Sci XXXIX-B1:469–473. doi:10.5194/isprsarchives-XXXIX-B1-469-2012

Yun M, Kimb J et al (2012) Application possibility of smartphone as payload for photogrammetric UAV system. Int Arch Photogramm Remote Sens Spat Inf Sci XXXIX-B4:349–352

Zhang W, Qi J et al (2016) An easy-to-use airborne LiDAR data filtering method based on cloth simulation. Remote Sens 8(6):501. doi:10.3390/rs8060501

Zhou G, Yang J et al (2012) Advances of flash LiDAR development onboard UAV. Int Arch Photogramm Remote Sens Spat Inf Sci XXXIX-B3:193–198

Chapter 5
Examples from the Boundaries of Geographic Survey: Architecture and Flood Modeling

Norbert Barkóczi, László Bertalan, Gergely Szabó, Márton Deák, Szabolcs Kari, Judit Csenge Vizi, Márk Zagorácz, András Sik, Miklós Riedel, Balázs Kohán, János Mészáros, Zoltán Szalai, Orsolya Szabó, and Balázs Nagy

Abstract This chapter will be more practical in nature. It will discuss two fields in which UAV-based photogrammetry proves a particularly efficient tool in geographic and architectural surveys. We will also reflect on the expectable accuracy of these relatively low-cost instruments.

N. Barkóczi • L. Bertalan • G. Szabó
University of Debrecen, Debrecen, Hungary
e-mail: norbert.barkoczi@gmail.com; bertalanlaszlo.geo@gmail.com; szabo.gergely@science.unideb.hu

M. Deák • S. Kari • J.C. Vizi • M. Zagorácz • A. Sik (✉) • M. Riedel
Lechner Tudásközpont Budapest, Hungary

B. Kohán
Eötvös Loránd University, Department of Environmental and Landscape Geography, Budapest, Hungary
e-mail: kohanb@caesar.elte.hu

J. Mészáros
Hungarian Academy of Sciences, Centre for Agricultural Research, Institute for Soil Sciences and Agricultural Chemistry, Department of Soil Mapping and Environmental Informatics, Tihany, Hungary

Z. Szalai
Eötvös Loránd University, Department of Environmental and Landscape Geography, Budapest, Hungary

Hungarian Academy of Sceinces, Geographical Institute, Research Centre for Astronomy and Earth Sciences, Budapest, Hungary

O. Szabó
Partium Christian University, Department of Economics, Oradea, Romania
e-mail: orsolya_szabo@yahoo.es

B. Nagy
Eötvös Loránd University, Department of Physical Geography, Budapest, Hungary

G. Casagrande et al. (eds.), *Small Flying Drones: Applications for Geographic Observation*, DOI 10.1007/978-3-319-66577-1_5

The first section will present a case study about the accuracy assessment of the digital stereophotogrammetry method (5.1). We will then present examples from two fields: the first one is architecture (5.2) and the second is flood modeling (5.3).

We give some examples about UAV applications in architecture because high-detail surveys focusing on one building are very different from low-detail surveys covering an entire settlement. The former are mostly used by architects and civil engineers; the latter belong more to the field of spatial planners and urban geographers. Flood modeling is mostly used by geomorphologists and disaster management experts and is more connected to physical geography.

These examples might be very different from a purely geographical point of view, but they present some methodological similarities. They provide an excellent base for comparing the criteria of different survey and flight planning techniques. In the case of building-scale architectural survey, the camera looks horizontally, and small details – in the order of centimeters or millimeters – must be captured. The output is a point cloud which will be later imported into an architectural software.

In the case of settlement-scale survey, the camera is looking at nadir, and 10–15 cm error is still not a big problem. The output is usually more GIS-related: this means that an orthophoto, a raster, or a digital elevation model is the absolute minimum. The point cloud is also a possible output which can be used, for example, to detect low-detail building models.

If we want to use a UAV for flood modeling, the acquisition method is similar to that of a settlement-wide survey: the camera points to the nadir and the covered area is relatively large; the main – and possibly only – output, however, is the digital terrain model.

The following sections focus on survey and data processing methods – mostly in point cloud format. The reader is therefore assumed to have some basic understanding of geographic survey and remote sensing in general.

Keywords UAV-based architectural survey • UAV-based topographical survey • Level of detail • UAV-based landscape monitoring • UAV-based risk management • Flood modeling

5.1 Case Study: Accuracy Assessment of the Digital Stereophotogrammetry Method

Norbert Barkóczi – László Bertalan – Gergely Szabó

The stereo digital photogrammetry (Egels and Kasser 2002) method can be used in many different areas and by many other disciplines (Linder 2009). Besides the lower-altitude flights with smaller UAVs (Tang et al. 2015), high-altitude satellite-based digital elevation model (DEM) investigations arise (Radhadevi et al. 2009) with potentially high accuracy results (Hobi and Ginzler 2012). The digital stereo-image interpretation can be a useful tool for natural hazard assessment (Pacheco 2003). Other studies published methods about creating point clouds and building

3D city models (Zhang et al. 2011). Several accuracy assessments have been carried out with different sensors and platforms (Van Leeuwen et al. 2009; Tampubolon and Reinhardt 2014).

The reliability and accuracy of the final results depend on several factors. Accuracy of DEMs had been compared based on different data sources (Rayburg et al. 2009). Wierzbicki et al. (2015) investigated how strong influence can the weather conditions have on the image quality. Earlier studies examined different interval sampling and flight heights (Li 1992). Comparative studies have been published before about digital terrain models based on various data models (Li 1994), and the uncertainties of DEMs have been defined (Weng 2002). Horizontal and vertical error assessment has been carried out before (Tang et al. 2015; Wierzbicki et al. 2015); however, in our case study, the difference between software processing levels and the air base distance between the images were investigated.

5.1.1 Tools and Study Area

A DJI Phantom 2 quadcopter was used for aerial mapping. The drone captured 414 images from the area of interest with a GoPro Hero 3+ Black Edition small action camera. These kinds of small digital cameras can produce submeter level accuracy (Ahmad 2011). The resolution of the images was 12 Mp. The average flight altitude was 75 m (~246 ft). The interval shooting time of the camera was set to 1 s. The average air base distance between the images was 10 m (~32.8 ft). For georeferencing eight ground control points (GCPs) were collected and additional 97 checkpoints were measured as reference data. The GCPs and the checkpoints were measured with an RTK-GPS (Stonex 9 – spatial accuracy ±2 cm). Three pairs of white plates were used for the GCP installation and one plate for the checkpoints, as they are all clearly visible and observable on the aerial images.

The study area is located in North-Eastern Hungary, near a small village named Basko. The average height above mean sea level was between 367 (~1,204 ft) and 426 m (~1,397 ft). The area is mainly used for grazing, so the average height of the vegetation was quite low; therefore, it was optimal for the photogrammetric survey. The eight GCPs and the additional checkpoints were distributed evenly in space.

5.1.2 The Photogrammetric Processing

For the photogrammetric processing, AgiSoft PhotoScan software was used (version 1.2.4). It provides an easy user interface and support lots of input and output data formats. After filtering the blurry or unwanted images, PhotoScan can load them into a project. The image preprocessing can affect the overall accuracy of the models (Ballabeni et al. 2015). The GoPro action camera has a wide-angle glass lens; however, the software can reduce this distortion effect (called fisheye). If the camera type is set to fisheye, then camera calibrations can be performed. Processing starts with the alignment of the photos into a spatial sequence. At this step the software calculates the camera positions and builds a sparse point cloud. The software provides five different quality levels in processing photo alignment and the buildup of the dense point cloud. These five parameter levels ("lowest," "low," "medium," "high," "ultrahigh") control the quality and the accuracy. The pair preselection was set to generic to find and detect the objects in the overlapping images. As we go from ultrahigh to the lowest accuracy setting, the software decreases the original resolution of the aerial photographs and downscales them (ultrahigh, no scaling; high, images are downscaled 2 times by each side; medium, downscaled 4 times; low, downscaled 8 times; lowest, downscaled 16 times). In this study due to the high memory requirements and the long processing time, the ultrahigh processing level had been omitted. In order to identify the GCP locations, both automatic and manual procedures are available. //PhotoScan supports guided marker placement, that is, a faster and efficient way to locate the GCPs. After the GCPs are located in each image and positions have been manually corrected, then these markers can be exported to use them in any future investigations. Once the GCPs have been located, their coordinates can be imported from a RTK-GPS text file. The file should contain the GCP names in it, matching with the GCP names in the PhotoScan, and their X, Y, and Z coordinates. The software calculates the relative errors of the GCP placement and, after the coordinate system is set, the absolute error of the GCPs as well. It is possible to run an optimization process in order to establish more accurate GCP locations, especially when accurate GCP coordinates are available. At the stage of building the dense point cloud, once again, five processing parameters are available. It is recommended to set the depth filtering mode to aggressive for aerial data processing. The DSM can be build based on the dense point cloud, and after that the orthophoto can be generated based on the DSM. If necessary the model boundaries and the model spatial resolution can be changed and be set to specified values. A report file can be generated after the photogrammetric processing is finished, and it contains the main parameters of the steps, processing level and times, errors, and number of the used images. Furthermore, PhotoScan supports batch processing: thus, we can set all the steps of the processing phases and create a general workflow. This opportunity helps when dealing with large datasets and long processing times. Also, the software is capable of running calculations via network-shared resources. If more computers connected to each other, then their performances can be used together as we create a cluster and configure a server and client computer(s) which can reduce the overall processing time.

5.1.3 Evaluations of Results

Different model scenarios were created depending on the applied images. Investigation was carried out to reveal how the accuracy of the models will change if we change the air base distance. The initial air base distance between two aerial

Table 5.1 Specifications of the presented model version

Number of images	Air base distance	Size of resolution reduction	DSM resolution (cm/pixel)	Ortho resolution (cm/pixel)	Point cloud density (points/m^2)	Tie points	Dense cloud points
414	10	1/4	12.8	3.2	62	65,496	12,111,335

Table 5.2 Descriptive statistics of the errors in meters

	X errors	Y errors	Z errors
No. of checkpoints	97	97	97
Minimum error	0.002	>0.000	>0.000
Maximum error	0.308	0.434	0.303
Mean	0.106	0.083	0.073
Std. error	0.007	0.008	0.007
Variance	0.004	0.006	0.004
Stand. dev	0.064	0.079	0.065
Median	0.093	0.062	0.059
25 percentile	0.057	0.026	0.026
75 percentile	0.147	0.121	0.100

images was 10 m (~32.8 ft). This distance was increased to 20 m (~65.6 ft), 50 m (~164 ft), and 100 m (~328 ft) to assess its effect on the models' accuracy. Furthermore, all models were processed with the four software processing parameters. In the end 16 different model scenarios were generated (4 different air base distance and 4 different processing levels). The observed (checkpoint measured with RTK-GPS) and predicted (DSM coordinates) values were extracted from the model for further statistical analysis. Descriptive statistics were calculated to evaluate each model version. For the statistical analysis, Past statistics software was used. The models were based on different number of images; therefore, the areas of the models are different. The processing parameter settings affect the final model size too. As expected, the overall error decreases as we increase the processing parameter from the 1/16 resolution reduction to 1/2 resolution reduction. In all cases the 1/16 resolution reduction models (lowest processing parameter) had the highest overall errors. The 1/8 resolution reduction models (low processing parameter) had lower overall errors, but the 1/4 and 1/2 resolution reduction models had the lowest overall errors. Between the 1/4 and 1/2 resolution reduction models, however, there was no significant difference.

From the 16 model versions, one model will be presented here in detail. The chosen model version had the lowest overall errors. Table 5.1 contains the specifications of the presented model version. The number of applied images was 414. The resolution reduction was set to 1/4 (medium processing parameter). The spatial resolution of the DSM was 12.8 cm and 3.2 cm for the orthophotograph. The point cloud density was 62 points/m^2.

Table 5.2 presents the descriptive statistics of the model. The evaluation of the models was carried out with 97 checkpoints. The table contains the horizontal (X,

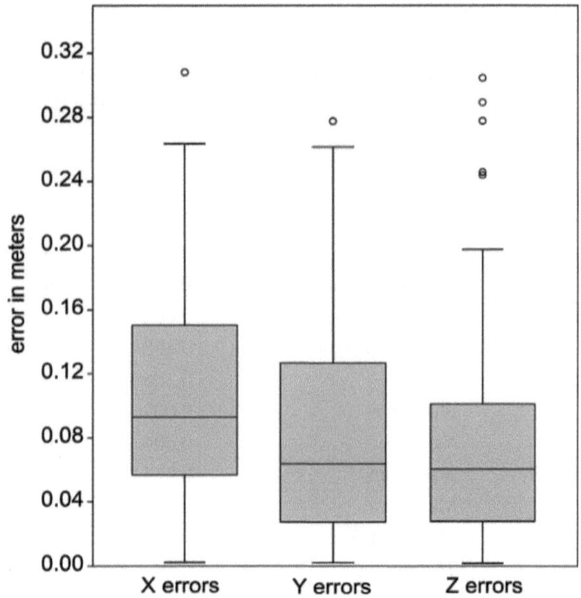

Fig. 5.1 Boxplot of the X, Y, and Z errors in meters

Y) and the vertical (Z) errors as well. The maximum errors were under 0.5 m in the Y direction and close to 0.3 m in X and Z directions. The mean average error (MAE) was 0.11 m in the X direction, 0.08 m in the Y direction, and 0.07 in the Z direction. The median value was lower than the MAE, as it can been seen in Fig. 5.1. The boxplot contains the distribution of the errors in each direction. The X direction has the highest median error (0.093 m), and there is a little difference between the Y median error (0.062 m) and the Z median error (0.059 m). The Z direction has the shortest interquartile distance between 0.1 and 0.026 m, which means that half of the vertical errors are in this range. Each direction has outlier values around 0.3 m; these values are outside of the 1.5 interquartile range. The standard deviation (SD) of the errors is between 0.064 and 0.079 m.

The spatial distribution of the GCPs and the checkpoints are presented in Fig. 5.2. The checkpoints were distributed equally in the study area just like the eight GCPs. The base map is the generated orthophotograph of the area. The size of the errors is presented with column bars. The red bars refers to the X errors, the green bars refers to the Y errors, while the blue ones for the Z direction errors. The most common column is the red, with X direction errors, although the highest error belongs to Y errors at the north side of the study area. At the center of the study area, the errors are lower than the errors at the north and southeast.

The optimal model version had an average 0.095 m horizontal error and 0.073 m vertical error. Although basically the lower resolution reduction means lower overall error, it is not necessarily desirable to raise the data processing parameter at the highest level. Furthermore, if enough image was captured during the flight, then it is possible to filter the photos and reduce the overlap, as the same accuracy still can be achieved, while the processing time can be greatly reduced. Generally the higher processing parameter results in higher accuracy, but there is not a significant differ-

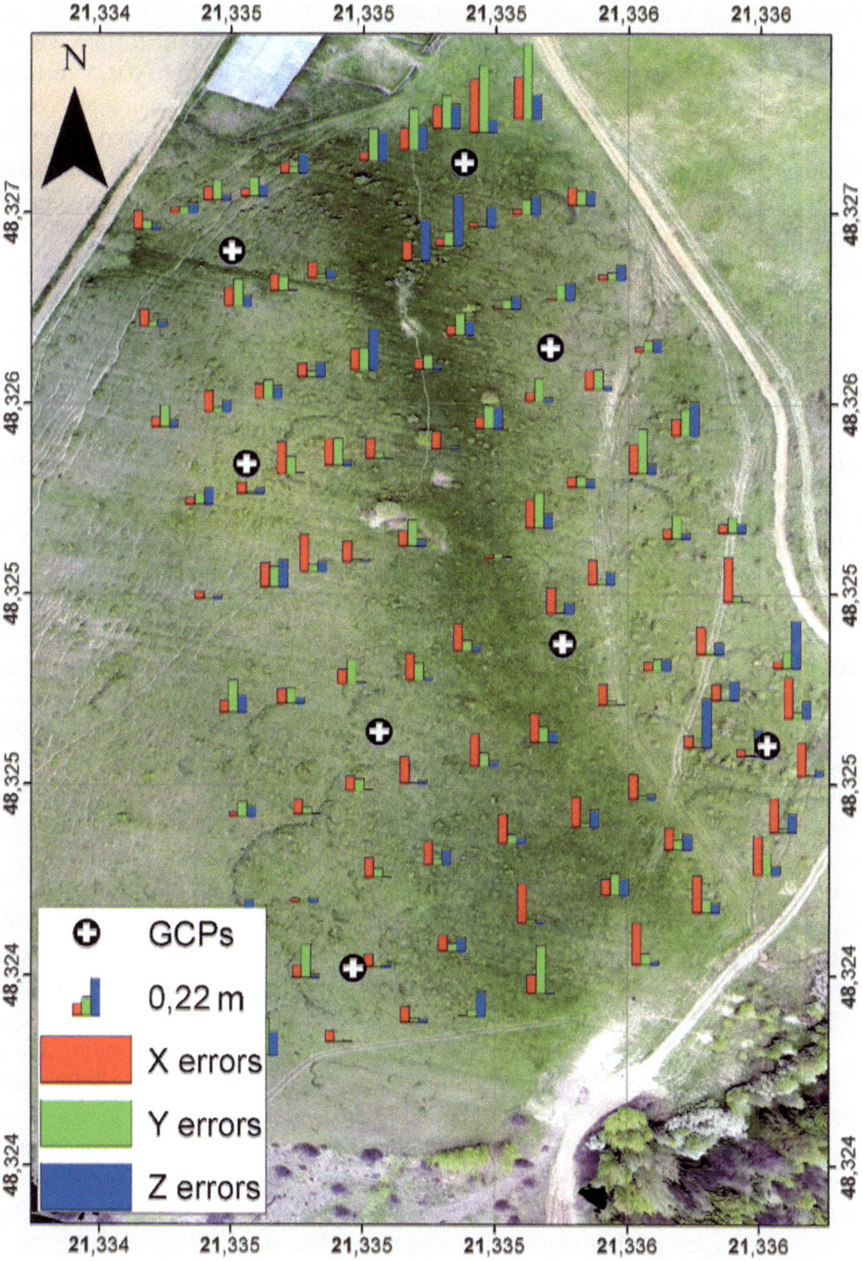

Fig. 5.2 The size of the errors at 97 checkpoints

ence between them, only in the processing time. However, the lowest parameter (1/16 resolution reduction) is only worth for a review phase.

Similar results in research articles and study cases were published in the topic. Barry and Coakley (2013) measured the field accuracy of this method with a use of

45 checkpoints and had a 0.023 m horizontal and 0.035 m vertical RMSE. Ahmad achieved a 0.623 m overall RMSE accuracy with similar small format action camera. Zhang et al. achieved a 0.027 m accuracy in X direction, 0.021 m in Y direction, and 0.08 m in Z direction without additional parameters using 80% endlap and 50% sidelap overlapping. Wierzbicki et al. (2015) compared different weather conditions and got an overall 0.11 m RMS error. These papers and our case study show that these systems and software can create reliable and accurate digital surface models and orthophotographs. The quick deployment and repeatability of these tools provide a useful system to researchers and entrepreneurs too.

5.2 Photogrammetric UAV Surveys in Architecture

Márton Deák – Szabolcs Kari – Judit Csenge Vizi – Márk Zagorácz – András Sik – Miklós Riedel

There are many cases where we can use a UAV in the field of architectural survey. A drone can be a supplement to other methods (such as laser scanners), but it is also possible that the whole survey is carried out by using a drone.

Our target can be a single building in high detail (Achille et al. 2015; Deák et al. 2017) and a whole town in low detail as well – essentially dividing the field of architecture into two subfields in this case: "building architecture" and "settlement architecture." At the individual building level, UAVs are basically a cheap and yet useful alternative for other mostly ground-based point cloud generating methods. These include terrestrial laser scanners (TLS) or mobile laser scanners (MLS). However, if the goal is to survey areas larger than about 2 km^2, a UAV with integrated cameras becomes a very good and affordable alternative to manned aircraft with high-resolution cameras.

In the following section, we will provide examples of both approaches, focusing on the mission planning and data processing steps. Aerial work described in Sect. 5.1 was performed by use of a DJI Phantom 2, that in Sects. 5.2 and 5.3 by a DJI Phantom 4 (Fig. 5.3). Processing was done by AgiSoft PhotoScan software.

Nevertheless, before rushing forward to the examples, it is worth spending a few words to clarify the basics of architectural survey and to emphasize that before any work can be started, a very careful planning is needed.

Since – as mentioned before – buildings can be modeled in high and low detail, we have to choose what we want. If we capture the target object in an unnecessarily high detail (e.g., the façade details in a building as part of a whole-city model), we have to expect a lot of extra processing time. Data storage may also become a serious issue. On the other hand, if we do not take into account the artistic details of an important building, we might have to go back and do the whole work all over again.

To avoid these complications, the following matters must be addressed: are details required in the output? If so, in what degree? How much data processing and

Fig. 5.3 DJI Phantom 4 UAV

storage capacity do we have? Which one is more important – details or the survey of a large area?

In order to efficiently answer these questions, we have to review a very important theme in architecture: the problem of different level of details (LoDs) and what can we expect from a UAV-based survey.

5.2.1 *Level of Detail*

Before every survey there has to be a planning phase where the exact details of the output are defined. To simplify this, architects around the world are working on LoD definitions, which are used for building modeling (Kari et al. 2016; Biljecki 2017, BIMforum 2016, Zagorácz et al. 2016). There are different approaches on how to classify a building and what detail means. It can be detail of information (in that case it is mostly called LoI as Level of Information), Level of Geometry (LoG), or combining them as Level of Development (LOD with capital letters). In this chapter we will use Level of Development strictly as a synonym to Level of Geometry.

The goal of predefined detail levels is simple. If one creates "LoD 2" data (be it a company, a research institute, or anybody else), they are incidentally stating many other information as well. The users will know what they can expect from the data and what data can or cannot be used for.

There are different types of classification. Architects designing buildings usually use "100-scale" LoD definitions, giving a certain "LoD" level for every building

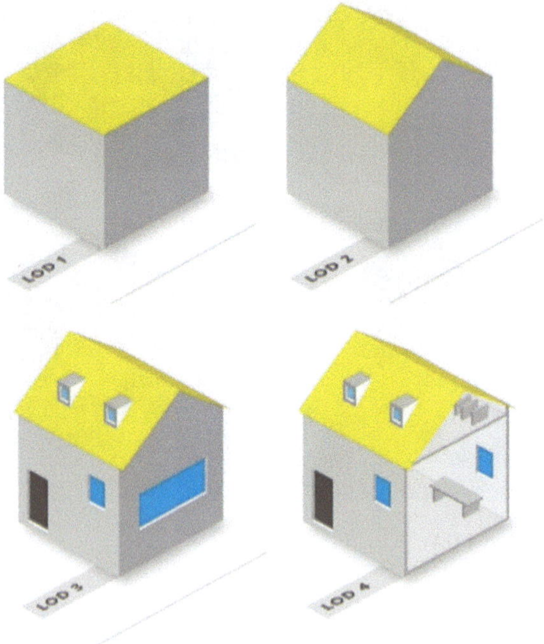

Fig. 5.4 LoD level definitions

element. It is important to note that it does not have to be the same level – a wall can be LoD 200, while an electric system could be LoD 400 (Volk et al. 2014).

The concept is somewhat the same if a plurality of buildings, forming a city, is considered. Just like an individual building, which includes different elements, a city consists of buildings. Different elements in a single building can be on different LoD levels, so buildings in a city model may be considered the same way.

Building LoD levels according to current professional conventions are marked from 0 to 4, meaning the following (Fig. 5.4):

- LoD 0: only the footprint of a building
- LoD 1: footprint extruded with the highest point of a building (effectively giving a cuboid)
- LoD 2: the plane of the walls and roofs – this is a simple building envelope
- LoD 3: façade details on each wall (doors, windows, balconies, and other objects) – effectively a detailed building envelope
- LoD 4: LoD 3 + details of inner objects (walls, rooms)

Creating a LoD 4 model by the mere use of UAVs might be very difficult, since indoor flight and survey is not practical for this purpose.

However from LoD 0 to LoD 3 UAV-based survey, it is very much possible.

Creating models is also not equally easy. This might surprise some, but given the way automation goes, creating a LoD 0 model (i.e., detecting the footprints) is

Fig. 5.5 The surveyed building

a rather tough task. Telling the difference between a building and another object (e.g., a tree, a car, a statue, or a rock) is not easy by looking at UAV-derived data. In most cases, we have to rely on human intuition. For any other automatically generated LoD models, LoD 0 is the base.

The next steps might be a bit complicated. The two cases below are examples of the creation of a LoD 3 model for an individual building, at expectable accuracy values, and the LoD 2 model of a larger area supplemented with other data.

5.2.2 *Building Scale Survey: Accuracy Assessment*

We tried to compare different building survey methods – focusing on the façades in this case. The study was based on the former Lágymányosi Tobacco Factory (which currently houses Lechner Knowledge Center – Fig. 5.5). In 2016 a survey of the building was started in order to preserve the architectural values of the building.

The survey was made using traditional methods such as measuring tape and laser rangefinder, supplemented by photographic documentation. We did a nondestructive survey on the building so the exploration of structural layers has not been done yet – and in this case, it is not needed as well (Figs. 5.6 and 5.7). However the valuable details were accurately assessed and documented for possible future renovations. The survey design documentation includes floor plans for each level, typical sections of the building, detailed drawings of the different façades, and also the schedule (a detailed list of specification of an architectural object type – for example, a window – which describes every property and additional info) of doors and windows.

In this case accuracy might be the big question: what advantages (if any) does UAV survey hold against other methods? In order to ascertain this, we also surveyed approximately 123 m^2 large façade using other methods as well (Fig. 5.8). We created two point clouds – one was obtained by a Leica P30 terrestrial laser scanner (TLS) and the other was the UAV photogrammetry-derived point cloud. The DJI Phantom 4 UAV captured 107 images, flying parallel to the façade for this purpose (Fig. 5.9).

Fig. 5.6 Render of the LoD 3 building model used for comparison

Fig. 5.7 Section of the hand surveyed LoD 4 model

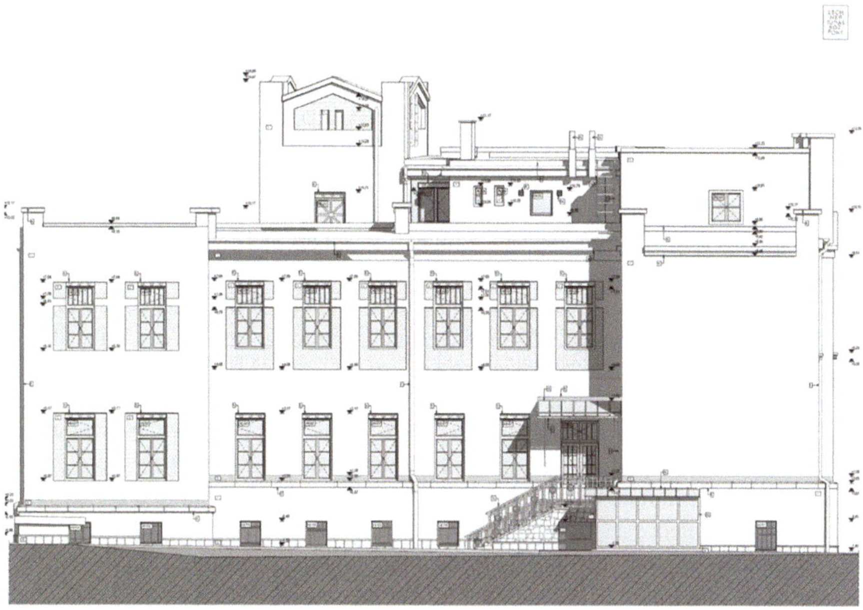

Fig. 5.8 Drawing of the façade used for comparison

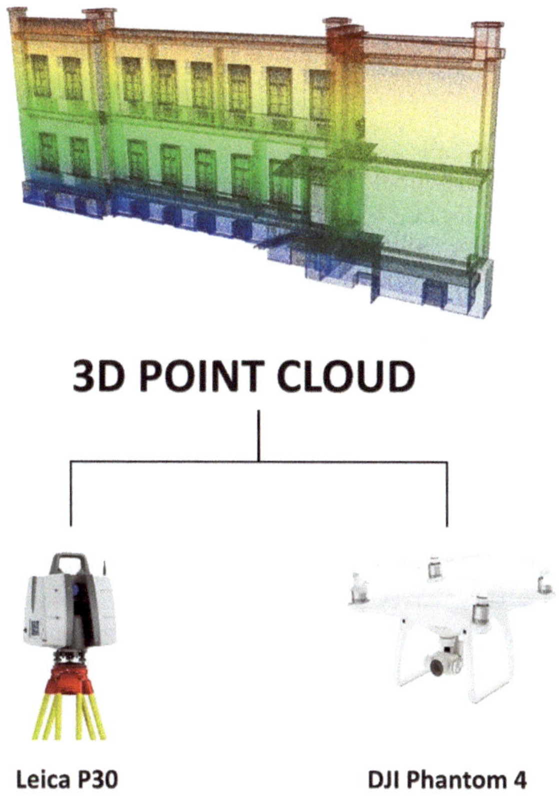

Fig. 5.9 The two instruments used for point cloud creation

For the achievement of high accuracy and point density, both vertical and horizontal overlaps between images were set to 80%. Flying distance from the façade was no more than 10 m, which is the closest the UAV could go without its obstacle avoidance system kicking in. On average, the same point was present in at least 4 images, but at most in 15. There were obviously exceptions – mostly at the edges – where the overlap was less than four images or when there was no overlap at all. Overall, the result was a very robust point cloud which was absolutely usable for architectural standards. However we also wanted to check the accuracy of the two methods (laser scanning and UAV-based photogrammetric survey) in comparison to the laser rangefinder-made building model.

As reference, a part of the southern façade was chosen, since there are many building elements such as corners, stairs, rails, doors, and windows. It is important to note that all comparisons are made with the model's idealized forms, meaning reality may be different, e.g., a wall surface is not completely flat or a corner is not exactly 90°-aligned.

The aim of the study was to determine if remote sensing methods (laser scanning and photogrammetry) produce similar results as a manual survey – and if not how different they are. We calculated the statistical descriptive parameters of the clouds and performed a detailed study on some building elements. This shows how accurate a UAV-based survey is and where is it better (or worse) than other methods. The idea in general is not new (Boehler and Marbs 2004; Grussenmeyer et al. 2008; Fritz et al. 2013), but its application in architecture is still not a widely covered field. We believe that this work might hold much new information for UAV pilots and architects alike.

Before any comparison is done, we have to lay down some foundations. If we compare two point clouds from an architectural perspective, the most important information is not the distance between the point clouds but also the difference in the derivable parameters. In other words, if we draw a building element (e.g., a window) based on two different data sources, the relevant thing is how differently it will be detected by either one. If the difference between the point cloud and the model is Δ, then the measured size of any building element (e.g., a window) using two different data sources cannot exceed 2Δ. However, in this case we should assume that the point cloud at the windows at short distances (within a window width) is mistaken in two opposite directions with maximum Δ value. The probability of this event almost equals to 0 due to the technical parameters of the instruments and procedures. From now on, then, we will assume that the overall accuracy of the point cloud is robust enough to allow for the comparison and the standard deviation in case if one building element does not exceed the value of the whole data.

In this case the UAV survey produced a more detailed result than the model which did not include some elements of the building that appeared on the point cloud (e.g., a hook for holding wires or a pipe). They were "hanging" from the model to about 30 cm, which was a strong influence on the maximum distance measured from the point cloud. They also affected the mean and the standard deviation values, but since the relative point number of these "non-modeled" elements is very small (0.2%), we considered its impact as nonsignificant.

Fig. 5.10 An example of the comparison between the model and the point clouds

Table 5.3 Statistical parameters of the distance between the two point clouds and the model

	Leica P30	DJI Phantom 4
Minimum (cm)	0	0
Maximum (cm)	25.630	28.498
Average (cm)	0.865	0.640
Standard deviation (cm)	2.361	2.127
Point density on the façade (p/m^2)	1,625	51,723

In the study we have calculated different statistical indicators, which can be interpreted on their own or can only be understood in relation to the reference. These are the point density (points/m^2 – i.e., the average number of points per m^2) and the descriptive statistics of the distance between each point in relation to the model (e.g., minimum, maximum, average and standard deviation) (Fig. 5.10 and Table 5.3).

Before the evaluation, both point clouds were accurately georeferenced and noise-filtered. It is also important to note that in the case of Leica P30, the point density depended on the distance between the instrument and the scanned object (i.e., the façade), so this value does not describe the characteristics of the instrument, but only the current sample. In this case, the distance between the building and the UAV was 10 m, while in the case of laser scanning, it was 50 m.

The photogrammetric point cloud was considerably more detailed. Even the pattern of powder plaster can be seen. The average distance between the model and the cloud was below 1 cm which makes it absolutely usable for high-detail surveys, e.g., for monumental buildings and statues. The Leica P30 was able produce similar results as well; however, it still represented the model less accurately than the photogrammetric survey. It is worth noting that the effective surveying speed of these two devices is not the same. The survey may take a few minutes with the Leica scanner, while surveying and data processing using a photogrammetric device, in this case a UAV, may be considerably more time-consuming.

Since numbers can hide certain errors, in addition to the statistical comparison, we also examined the individual building elements. A typical issue of the photogrammetric dataset was that the flat wall surfaces were well covered, but it produced more rough results at the corners. The difference is most noticeable in the case of vertical rods of iron railing where the static scanner provided a much more detailed

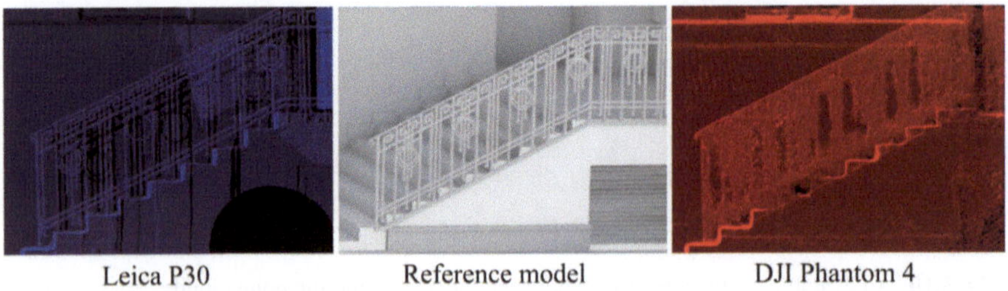

Fig. 5.11 Difference of the two point clouds in case of the railing

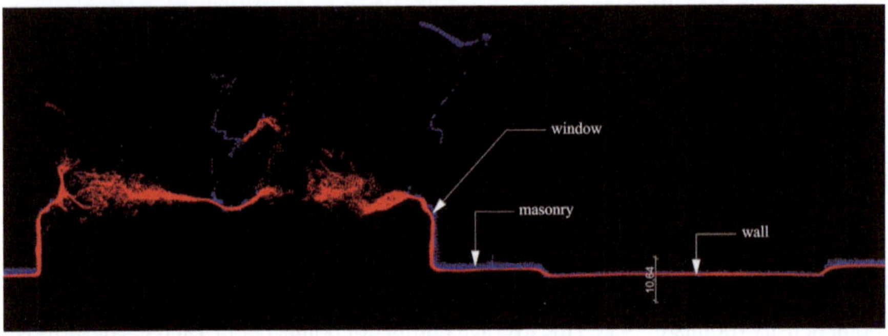

Fig. 5.12 The difference of some building elements

data, but in case of the photogrammetric point cloud, many iron rods of the railing appeared as one object (Figs. 5.11 and 5.12).

This does not mean that UAV-derived photogrammetric data would be useless. In fact, it was more precise and accurate in several cases. Nevertheless, the user should take these errors into the proper account. Our conclusion is that a UAV-based photogrammetric point cloud might be more accurate in general, but in case of small details, it fails against the terrestrial laser scanners.

Therefore we think it is recommended to supplement static scanning measurements by a UAV photogrammetric survey.

5.2.3 Building Scale Survey: A Demonstrational Survey

While the achievement of high accuracy is possible – as seen above – it might take a lot of work. In the previous case, 107 photos for a 123 m^2 large façade were taken, meaning a whole building can be covered at least with 3,000–4,000 images (depending on the details as well). In the next example, we created two models. Our goal was not architectural survey (at least not in the documenting and planning sense) but 3D data demonstration.

Fig. 5.13 Central building of the Hungarian Academy of Sciences

Knowing the exact expectable accuracy parameters of the output point cloud, we selected another, absolutely monumental building with a lot of detail on the façade. We chose the central building of the Hungarian Academy of Sciences (HAS) which was inaugurated in 1865 and is one of the prominent landmarks in Budapest (Fig. 5.13).

During this work we created two different models. One is a textured mesh (Fig. 5.14) and the other is a detailed point cloud of the main façade (Figs. 5.15 and 5.16).[1] The mesh is made solely for demonstrational purposes – to show the building to those who can't see it in person or to look at it from angles which are physically not possible to obtain. The point cloud however had a different purpose. In case the artistic details of the statues or the building needed to be renovated or presented in detail to the public, this data might prove to be invaluable.

Meshing and model creation are somewhat out of the scope book. Still, we believe it is essential to emphasize what we couldn't do without a drone. A survey – either by photogrammetry or laser scanning – is absolutely possible without a drone. We can use terrestrial instruments to create 3D models – and from a defined point of view, the difference won't be even that much relevant. However we have to consider that a

[1]The detailed point cloud of the central building of the Hungarian Academy of Sciences can be reached here: http://webmap.lechnerkozpont.hu/webappbuilder/apps/MTA2/.

Fig. 5.14 Low-resolution model of the HAS central building

Fig. 5.15 3D point cloud of the main façade

Fig. 5.16 A detailed view of the point cloud

Fig. 5.17 Point cloud of the inner city of Tállya

UAV can fly in façade at all levels, taking perpendicular images even from the highest parts of the building. When looking from the ground, a lot of details are hidden, obscured by ledges or other elements. If we can collect photogrammetric data from above, these covered areas tend to disappear.

In some cases detailed surveying needed to be done. A TLS cannot collect high enough point density to be able to model such parts as the Corinthian columns and statues – these elements are often scanned using photogrammetry during every survey. In this case however the statues are on a high level on the façade which basically denies us the application of any handheld photogrammetric instruments. Therefore only a UAV could fly close enough to these parts and take the photos needed for processing.

In conclusion, we strongly advise that high-detail survey of façades should always be supplemented by UAV survey.

5.2.4 City-Scale Survey

At first sight, a town-scale project might not be ideal for a UAV. This is wrong impression, though. Due to their low cost, UAVs can be competitive in this field and might replace some aerial surveys carried out by small airplanes.

One example is surveying project of the Hungarian town of Tállya. We conducted a survey producing a point cloud (Figs. 5.17 and 5.18), a DEM (Fig. 5.19), an orthophoto (Fig. 5.20), and LoD 2 building models (Fig. 5.21). The town itself is

Fig. 5.18 A more detailed view of the point cloud

Fig. 5.19 An example of the DSM

relatively small (the built-up area is 1,5 km^2) and the population of Tállya is 1892 as of 2015 (TeIR-KSH).

This is maybe the largest area we can survey using a small UAV like the Phantom 4. Processing the vast amount of data (around 5,000 images were captured) might be too much for a personal computer or even for a modern workstation. Processing time can take up to 400 h (pure computing time) on a computer with 16GB RAM, i7–4770 CPU, and an NVIDIA Quadro K2000 GPU with 10GB VRAM. It is a very

Fig. 5.20 Digitized building footprints on the orthophoto

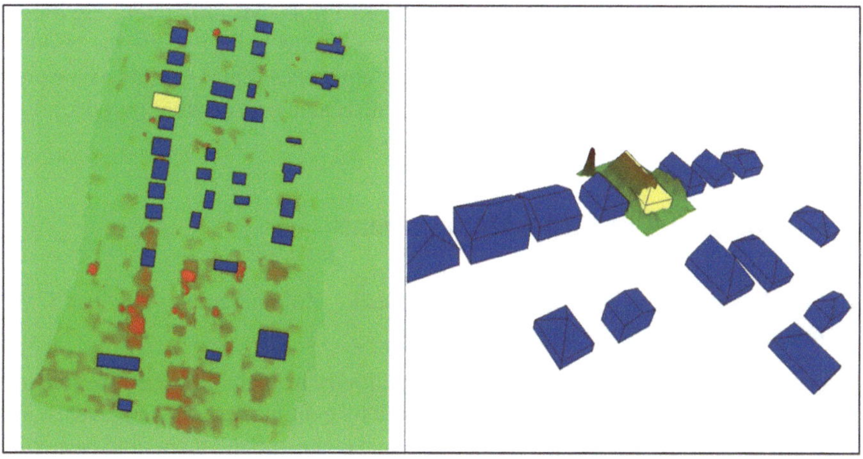

Fig. 5.21 LOD2 building models

time-consuming task, and it also requires serious financial investments regarding the IT infrastructure (Verhoeven 2011; Turner et al. 2014).

The primary purpose of the survey was to produce digital data for the local government of Tállya, who could use it for administrative, spatial planning, and general municipal tasks. These data can be useful inputs during tasks such as building control procedures, mapping, and planning workflows. A UAV-derived orthophoto allows us to create a current, accurate, and detailed dataset. It displays the status of public objects (such as street furniture) based on the exact coordinate position of the artificial environment.

To create an orthophoto, we need to create a 3D model first (DEM for a traditional orthophoto, DSM for a true ortho). The first step was to evaluate 3D images. We produced the point cloud of the settlement by the aforementioned AgiSoft PhotoScan Software. To reduce run time, not all steps have been taken with the highest precision settings. For accurate orthophotos, it was necessary to conduct the most accurate possible camera calibration, but in order to reduce time "dense matching," it was only done in medium quality.

The area was covered during 12 flights (1 flight is 1 battery duration) and the height of each flights was 100 m above ground level. The output resolution of the DEM became 7.6 cm/px, while the orthophoto was 4.5 cm/px. The resulting 3D point cloud was used later for LoD 2 modeling, as in the side walls and planes were partly detected manually and partly automatically. As we already mentioned above, extracting building footprints is a generally time-consuming manual process. In this case we used the output orthophoto to digitize them.

For ground classification we used the Statistical Cloth Filter algorithm (Zhang et al. 2016), so we could interpolate solid, raster models – to be exact digital terrain model (DTM) using only the ground points and Digital Surface Model using all of the points. We then subtracted the DTM from the DSM, effectively creating a normalized digital surface model (nDSM). The nDSM values inside the building footprint basically gave us the LoD 2 building envelopes. Later on, this data can be suitable for facility management, 3D visualization analysis, or other purposes.

5.3 Challenges of Flood Management

Balázs Kohán – János Mészáros – Márton Deák – Zoltán Szalai – Orsolya Szabó – Balázs Nagy

UAV surveys can help to better understand floods in their different phenomenal causes and characters. Floods may be triggered by extreme precipitation events, and their effects are also dependent on environmental context. Settlements and properties can be seriously endangered by inland inundations, flash floods in hilly or mountainous areas. Adequate protection against this kind of environmental risk can be achieved only by the implementation of appropriate decision preparation and decision support methods. The development of those implies, on the one hand, the availability of accurate preventive survey systems; on the other hand, it requires the possibility of frequently reiterating observations through sufficiently affordable tools.

National risk mapping programs were carried out in Hungary by the General Directorate of Water Management and by NAGIS (National Adaptation Geoinformation System); they brought the creation of a database. The expectation is that such information would prove suitable for supporting decision-making. However, both the aforementioned projects are based on low-resolution data (e.g., relief models); thus, they cannot be effectively used in high-resolution risk analysis, at individual property level.

According to the results of the latest model-based forecasts, toward the end of the twenty-first century, the Carpathian Basin will probably be affected by precipitation conditions that are even more extreme than those of today. According to the estimations, summers will be characterized by longer dry periods (Pongrácz et al. 2014), while precipitations exceeding 20 mms will occur more scarcely, but in a more intensive way. The intensity of the precipitation will expectedly increase with each season. In the winter and fall seasons, the frequency and intensity of the extreme precipitation events will increase (Bartholy et al. 2015); thus, more and more flood waves can be expected.

As far as we know, up to this date, there has not been any underpinning research carried out with the purpose of developing a cheap model to analyze floods and inland inundations using UAV devices.

In Hungary the researchers of the University of Debrecen have developed out relief models using drones (Szabó and Mecser 2014), in order to monitor the changes in the riverbed of the Sajó river (Bertalan and Szabó 2015) or to determine the ideal position of solar cells (Szabó et al. 2015). We can also find international examples, which describe well the methodology in general (Koutsoudis et al. 2013). Other water-related applications can cover the field of geomorphological erosion – which is also a factor to be considered in the event of flooding (Mészáros et al. 2016).

The goal was to be achieved almost solely using a DJI Phantom 2 UAV, with a GoPro Hero 3+ Silver edition camera, capable of taking 10 MPx images.

Toward flood prevention, it is essential to be able to foresee the movement of water on floodplains, the potential dimension of the flood, and the flood-related risk.

This type of knowledge requires a more detailed information and awareness about the topography of floodplains and the spatial connections among them, the floodless areas and the riverbeds.

The needed information can be obtained by means of surveys and analyses so that the potential flood path and the flood extension can be appropriately identified.

Such work requires a fast, precise, flexible, and cost-effective set of tools and techniques, whose deployment can be repeated as needed in order to obtain updated information.

In the areas of inhabited embanked floodplains, the increasing flood levels are especially problematic if floodplains and floodless areas (e.g., river terraces) are not sufficiently separated. These areas are often connected to each other with small differences in altitude of the different levels, and the relief has become even due to artificial soil recharge and artificial soil alignment. Floodplains and the surroundings, furthermore, have often become built environments. The situation is aggravated by the floodplain filling caused by the repeated floods. Thus the potentially flooded areas cannot be defined by the traditional use of maps or by the analysis of aerial photos or space images or by field visits or by the measurement of different points of the area. The solution lies in the full coverage, high-resolution mapping of the floodplain areas and the adjacent floodless areas, based on which inundation models can be developed. Models allow to predict, at individual property level, the extent of inundations caused by current floodplain configuration. Since the relief of

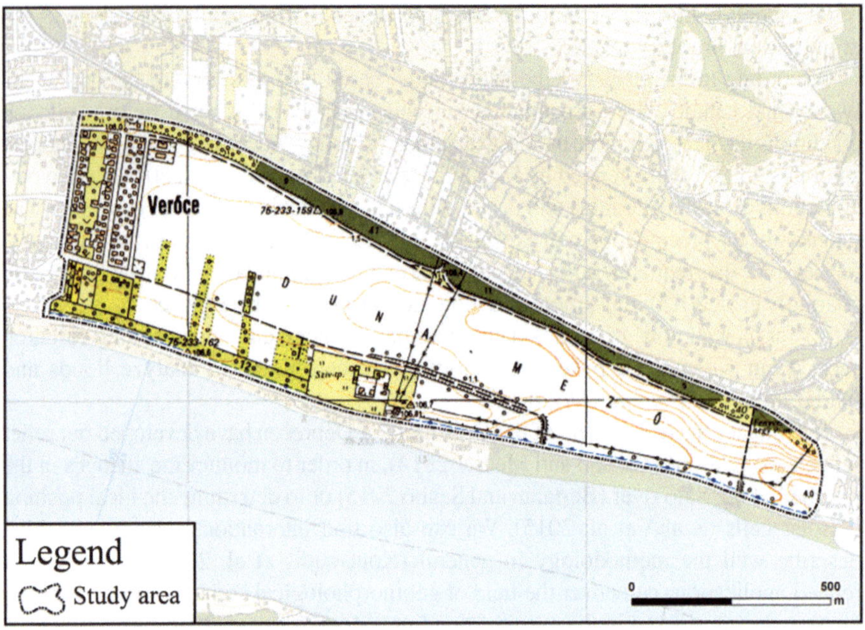

Fig. 5.22 The study area

the floodplains is modified by each flood, we should choose a method that besides its needed accuracy can be repeated after each flood period.

This way, the long-term assessment of inundation phenomena can be carried out: the protection against floods becomes more efficient and the installation of the preventive measures is possible. We wanted to analyze an area of 2–3 km^2, by developing a relief model of a 2–3 cm accuracy. This was to be based on photos taken by a camera-equipped UAV from approximately 80 m a.g.l. The survey provided for an adequate basis to prepare the inundation analysis and the flood forecasts. It can even be used to create a flood emergency plan. The method can also be used if an inland inundation simulation is needed.

The goals of the research project were as follows:

- Simulation of floodplain and inland inundations using a newly created relief model and taking into consideration the level of river waters
- Detection of the path of all floods which could endanger populated areas, evaluating the inundations in populated areas, taking into consideration the water levels
- Definition the water depths associated with flood inundations in the populated areas:
  - Creation of a fast, cost-efficient methodology to be widely used
  - Test of such methodology based on previously occurred, well-documented inundations

Fig. 5.23 Maximum water level in 2013 (June) in the Duna-mező (Image: Selmeczi Kovács Ádám)

The study area is the Verőce region, lying in the Danube Blend, on the left bank of the Danube. East to Verőce settlement lies the 2 km long Duna-mező (Duna field) floodplain (Fig. 5.22).

The northern border of the floodplain is the 6–10 m high railway embankment. On the southern border of the floodplain, lying close to the Danube, is road no. 12, which was inundated several times in the recent past (2002, 2006, 2010, 2013). The road and its foundation is not an embankment: it does not function as a dyke for the flooded Danube. Moreover it is officially considered and marked as a road with increased flood risk (our floodplain inundation simulation model does not comprise road no. 12, as the study area lies north to this road) (Fig. 5.23).

If we want to analyze an area of 2–3 km^2; the most cost-efficient method is to create a relief model based on the photogrammetric analysis of a photos taken by a camera-equipped UAV. The aim of this method is to process images taken from different angles to reconstruct the shape of shown object by using the parallax difference between them. Using this method we created the photogrammetric point cloud of the area, classified the data, and created a DEM for further analysis.

The point cloud was transformed to EOV/HD72 projected coordinate system using ground control points so that it could be processed in a GIS environment. In order to achieve this, easily distinguishable components of the terrain were identified (the corner of the fence, a light pole, edge of the railway embankment, etc.) and then the 3D coordinates of the aforementioned GCP system were recorded using an RTK-GPS.

A further step of the data processing is to identify and remove the errors occurring in the point cloud. This was to be carried out using the Statistical Outlier Filter algorithm of the freely accessible CloudCompare software. The relief model was created by the interpolation of the reduced point cloud using the Inverse Distance Weighting algorithm. The further hydrologic analysis and the representation of the

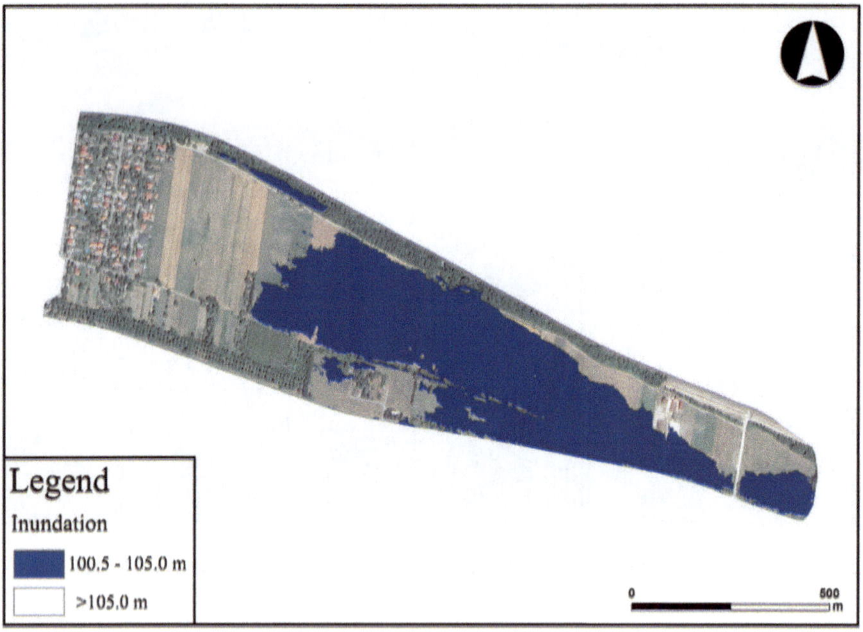

Fig. 5.24 The extent of the flooding at water level of 105 m AMSL

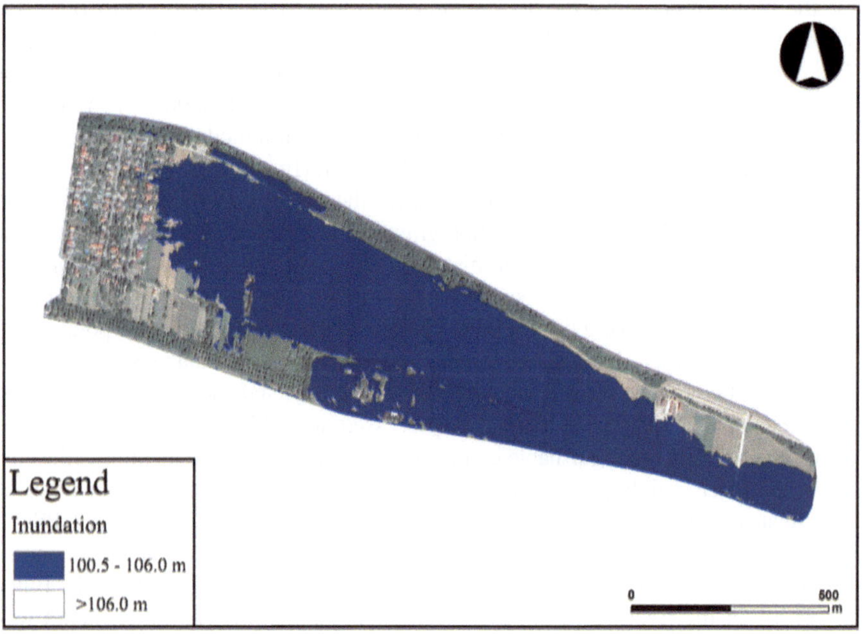

Fig. 5.25 The extent of the flooding at water level of 106 m AMSL

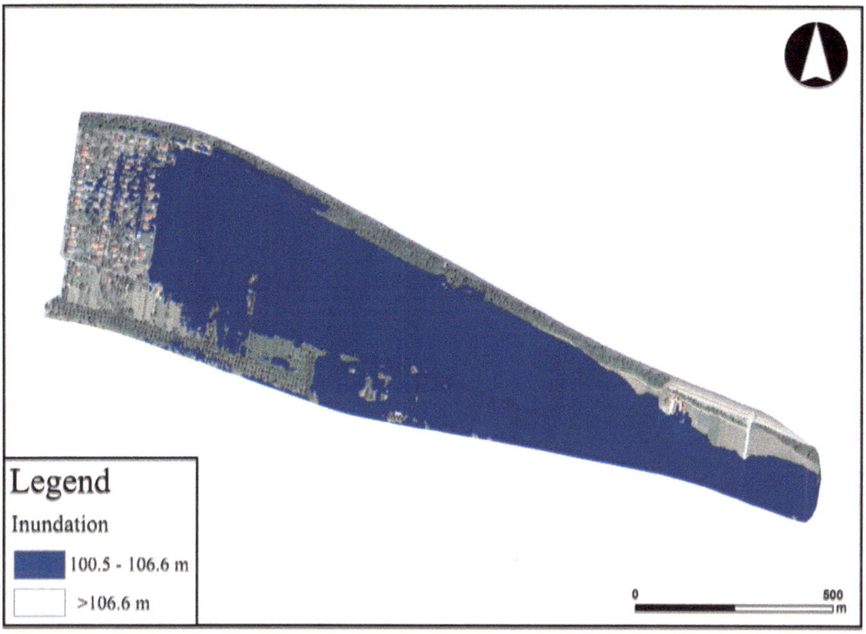

Fig. 5.26 The extent of the flooding at water level of 106.6 m AMSL

flood levels could be carried out either by ArcGIS 10.3 or by the freely accessible Quantum GIS. We also created an orthophoto to help us interpreting the inundation levels (Figs. 5.24, 5.25, and 5.26).

As Verőce does not have an official measuring station, the flood levels were deducted using the data provided by the measuring stations of Nagymaros and Vác. The highest flood level in 2013 was 751 cm at Nagymaros and 804 cm at Vác, which in absolute above sea level means 106,94 AMSL and 106,16 AMSL flood level. The linear interpolation estimation using these two values resulted in approximately 106,6 AMSL at Verőce. The calibration of the model was carried out using the aerial photographs taken at the peak of the flood from 2013.

The pictures show the real and the estimated flooding areas basically cover each other. When the maximal flooding was analyzed, we could see that the real flooding and the estimated flooding corresponded at an individual property level. Maps were created to illustrate the different flood levels and flooding animations were carried out. Flooding depths were estimated by subtracting the surface that resulted from the photogrammetric processing from the existing water level height data (Fig. 5.27).

According to the results of our estimation, the water level of the Danube has to be 102.24 m high AMSL so that water would appear at the lowest point of the Duna-mező. At this level at the measuring point at Nagymaros, the water level was around 311 cm (±5 cm). At an absolute water level of 105 m, the flooding reaches the first

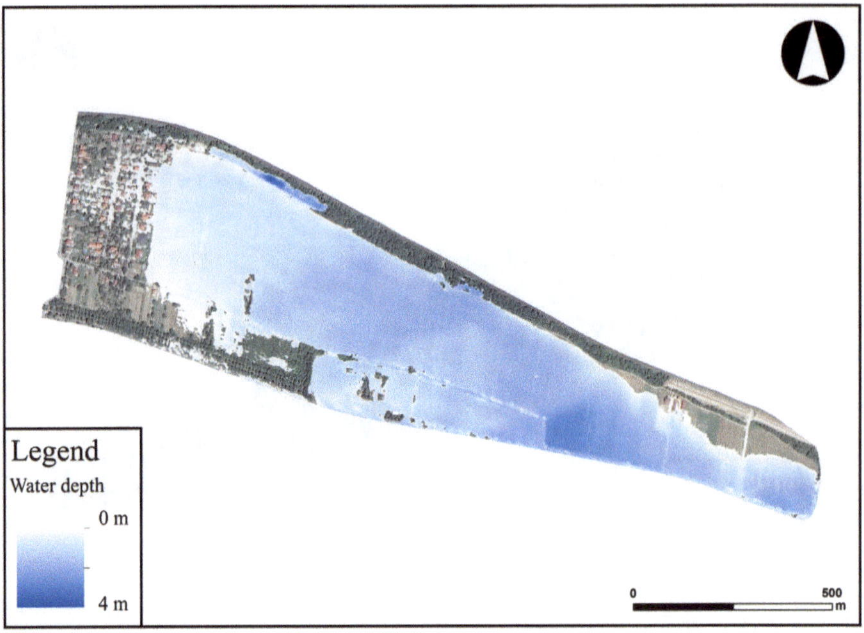

Fig. 5.27 The depth of the flooding at a water level of 106,6 m (mBf)

house of the settlement and water appears in the basements of the low-lying houses of the eastern part of the village. At 105,74 AMSL, the water reaches the houses of the first street, the Dunasor. At 106 AMSL the gardens of the low-lying properties at the edge of the settlement are under water. If the water level would increase with 20 cm, the flooding reaches the middle of the street and water appears in the second street of the settlement (Váci Mihály Street) too.

References

Achille C, Adami A et al (2015) UAV-based photogrammetry and integrated technologies for architectural applications – methodological strategies for the after-quake survey of vertical structures in mantua (Italy). Sensors 15(7):15520–15539. doi:10.3390/s150715520

Ahmad A (2011) Digital mapping using low altitude UAV. Pertanika J Sci Technol 19(S):51–58

Ballabeni A, Apollonio FI et al (2015) Advances in image pre-processing to improve automated 3D reconstruction. The International Archives of the Photogrammetry. Remote Sens Spat Inf Sci XL-5(W4):315–323. doi:10.5194/isprsarchives-XL-5-W4-315-2015

Barry P, Coakley R (2013) Field accuracy test of RPAS photogrammetry. Int Arch Photogramm Remote Sens Spat Inf Sci XL-1(W2):27–31

Bartholy J, Pongrácz R, Kis A (2015) Projected changes of extreme precipitation using multi-model approach. Időjárás 119(2):129–142

Bertalan L, Szabó G (2015) Lateral erosion monitoring along a southern section of (Slaná) river. In: Křížova A (ed) Detailed aerial mapping and flood impact monitoring in the V4 region. University Faculty of Mathematics and Physics, Bratislava

BIMForum (2016) Level of development (LOD) specification. Available via http://bimforum.org/lod/. Accessed 8 May 2017

Biljecki F (2017) Level of detail in 3D city models. PhD Thesis, TU Delf. doi:10.4233/uuid:f12931b7-5113-47ef-bfd4-688aae3be248

Boehler W, Marbs A (2004) 3D scanning and photogrammetry for heritage recording: a comparison. Geoinformatics 2004:291–298

Deák M, Kari Sz et al (2017) Novel façade surveying method: comparison of photogrammetry, static and mobile laser scanning. Metszet, in press (in Hungarian)

Egels Y, Kasser M (2002) Digital photogrammetry. Taylor & Francis, London

Fritz A, Kattenborn T, Koch B (2013) UAV-based photogrammetric point clouds – tree stem mapping in open stands in comparison to terrestrial laser scanner point clouds. International Archives of the Photogrammetry. Remote Sens Spat Inf Sci XL-1(W2):141–146

Grussenmeyer P, Landes T et al (2008) Comparison methods of terrestrial laser scanning, photogrammetry and tacheometry data for recording of cultural heritage buildings. Int Arch Photogramm Remote Sens Spat Inf Sci XXXVII Part B5 37:213–218

Hobi ML, Ginzler C (2012) Accuracy assessment of digital surface models based on WorldView-2 and ADS80 stereo remote sensing data. Sensors 12(5):6347–6368. doi:10.3390/s120506347

Kari Sz , Lellei L et al (2016) BIM to GIS and GIS to BIM. Paper presented at the CAADence in Architecture. 16–17 June 2016, Budapest, Hungary, p 67–72. doi: 10.3311/CAADence.1645

Koutsoudis A, Vidmar B et al (2013) Multi-image 3D reconstruction data evaluation. J Cult Herit 15(1):73–79. doi:10.1016/j.culher.2012.12.003

Li Z (1992) Variation of the accuracy of digital terrain models with sampling interval. Photogramm Rec 14(79):113–128. doi:10.1111/j.1477-9730.1992.tb00211.x

Li Z (1994) Comparative study of the accuracy of digital terrain models (DTMs) based on various data models. ISPRS J Photogramm Remote Sens 49(1):2–11. doi:10.1016/0924-2716(94)90051-5

Linder W (2009) Digital photogrammetry a practical course. Springer Science & Business Media, Berlin

Mészáros J, Árvai M et al (2016) UAV based 3D digital surface model to estimate paleolandscape in high mountainous environment. Geophys Res Abstr 18: 15430. Paper presented at the EGU General Assembly 2016, 17–22 April, 2016 in Vienna, Austria

Pacheco JA (2003) Digital stereo image interpretation for natural hazard assessment. Thesis, International Institute for Geo-Information Science and Earth Observation

Pongrácz R, Bartholy J, Kis A (2014) Estimation of future precipitation conditions for Hungary with special focus on dry periods. Időjárás 118(4):305–321

Radhadevi PV, Jacobsen K et al (2009) Comparison of digital elevation models generated from SPOT-5 HRS stereo data and CARTOSAT-1 stereo data. Paper presented at the ISPRS Hannover Workshop 2009, 'High-Resolution Earth Imaging for Geospatial Information, p 1–5

Rayburg S, Thoms MC, Neave M (2009) A comparison of digital elevation models generated from different data sources. Geomorphology 106(1–3):261–270. doi:10.1016/j.geomorph.2008.11.007

Szabó G, Mecser N (2014) New tecnhique in spatial data collection-low cost DSM and ortophoto generating with drone. In: Boglárka B (ed) Az elmélet és a gyakorlat találkozása a térinformatikában V: Térinformatikai konferencia és szakkiállítás. Available via http://geogis.detek.unideb.hu/TKonferencia/2014/Konferenciakotet_2014.pdf. Accessed 8 May 2017 (in Hungarian)

Szabó S, Enyedi P, Horváth M, Kovács Z, Burai P, Csoknyai T, Szabó G (2015) Automated registration of potential locations for solar energy production with light detection and ranging (LiDAR) and small format photogrammetry. J Clean Prod 112(5):3820–3829. doi:10.1016/j.jclepro.2015.07.117

Tang S, Liu Y et al (2015) Research on the processing technology of low-altitude unmanned aerial vehicle images. Paper presented at the Proceedings of the SPIE 9808, International Conference on Intelligent Earth Observing and Applications 2015, 980830 (December 9, 2015). doi: 10.1117/12.2207583

Tampubolon W, Reinhardt W (2014) UAV data processing for large scale topographical mapping. Int Arch Photogramm Remote Sens Spat Inf Sci XL-5:565–572. doi:10.5194/isprsarchives-XL-5-565-2014

TeIR-KSH. Available via https://www.teir.hu/helyzet-ter-kep/kivalasztott-mutatok.html?xteiralk=htk&xids=1001,1002&xtertip=T&xterkod=1221 (in Hungarian). Accessed 8 May 2017

Turner D, Lucieer A, Wallace L (2014) Direct georeferencing of ultrahigh-resolution UAV imagery. IEEE Trans Geosci Remote Sens 52(5):2738–2745. doi:10.1109/TGRS.2013.2265295

Van Leeuwen B, Tobak Z et al (2009) Small format aerial photography – a cost effective approach for visible, near Infrared and thermal digital imaging. In: G. G. Car A. (ed) Geospatial crossroads @ GI_Forum' 09. Heidelberg: Proceedings of the Geoinformatics Forum Salzburg, p 200–209

Verhoeven G (2011) Software review: taking computer vision aloft – archaeological three-dimensional reconstructions from aerial photographs with PhotoScan. Archaeol Prospect 18(1):67–73. doi:10.1002/arp.399

Volk R, Stengel J, Schultmann F (2014) Building information models (BIM) for existing buildings – literature review and future needs. Autom Constr 38:109–127. doi:10.1016/j.autcon.2013.10.023

Weng Q (2002) Quantifying uncertainty of digital elevation models derived from topographic maps. In: Richardson DE, van Oosterom P (eds) Advances in spatial data handling. Springer, Berlin, pp 403–418

Wierzbicki D, Kedzierski M, Fryskowska A (2015) Assessment of the influence of UAV image quality on the orthophoto production. The International Archives of the Photogrammetry. Remote Sens Spat Inf Sci XL-1(W4):873–877. doi:10.5194/isprsarchives-XL-1-W4-1-2015

Zagorácz M, Etlinger J, Iványi P (2013) Accurate quantity data extraction from low level of detail BIM models. Paper presented at the Conference: The Fourteenth International Conference on Civil, Structural and Environmental Engineering. doi: 10.4203/ccp.102.155

Zhang W, Qi J et al (2016) An easy-to-use airborne LiDAR data filtering method based on cloth simulation. Remote Sens 6(8):501. doi:10.3390/rs8060501

Zhang Y, Xiong J, Hao L (2011) Photogrammetric processing of low-altitude images acquired by unpiloted aerial vehicles. Photogramm Rec 26(134):190–211. doi:10.1111/j.1477-9730.2011.00641.x

Index

G. Casagrande et al. (eds.), *Small Flying Drones: Applications for Geographic Observation*, DOI 10.1007/978-3-319-66577-1